T0255446

CAMBRIDGE LIBRARY COLLECTION

Books of enduring scholarly value

Monographs of the Palaeontographical Society

The Palaeontographical Society was established in 1847, and is the oldest Society devoted to study of palaeontology worldwide. Its primary role is to promote the description and illustration of the British fossil flora and fauna, via publication of an authoritative monograph series. These monographs cover a wide range of taxonomic groups, from microfossils, trilobites and ammonites through to Coal Measure plants, mammals and reptiles, and from all ages from Cambrian to Pleistocene. They form a benchmark for understanding the past life of the British Isles and many include the original descriptions of numerous key species. The first monograph (on the Crag Mollusca) was published in March 1848 and the Society still continues this work today. Notable authors in the series include Charles Darwin (fossil barnacles) and Richard Owen (dinosaurs and other extinct reptiles). Beginning in 2014, the Cambridge Library Collection and the Society are collaborating to reissue the earlier publications, focusing on monographs completed between 1848 and 1918.

A Monograph of the British Carboniferous Trilobites

Henry Woodward (1832–1921) intended this monograph, originally published as two parts in 1883–4, to be the first of a series of supplements aimed at completing Salter's unfinished work on British trilobites. In the event, no others were published. As the first monographic treatment of Carboniferous trilobites from Britain and Ireland, it has remained a standard work. Thirty-one species distributed among four genera are described, including the type species of the well-known *Phillipsia*, *Griffithides* and *Brachymetopus*. The specimens are faithfully reproduced in a series of fine illustrations. The monograph concludes with three short appendices: the nature of pores on the trilobite headshield; a new Carboniferous trilobite from Ohio; and a letter from Agassiz on possible affinities between trilobites and a marine isopod. Woodward described further new species of Carboniferous trilobites in a series of papers published between 1884 and 1906, which acted as supplements to his monograph.

Cambridge University Press has long been a pioneer in the reissuing of out-of-print titles from its own backlist, producing digital reprints of books that are still sought after by scholars and students but could not be reprinted economically using traditional technology. The Cambridge Library Collection extends this activity to a wider range of books which are still of importance to researchers and professionals, either for the source material they contain, or as landmarks in the history of their academic discipline.

Drawing from the world-renowned collections in the Cambridge University Library and other partner libraries, and guided by the advice of experts in each subject area, Cambridge University Press is using state-of-the-art scanning machines in its own Printing House to capture the content of each book selected for inclusion. The files are processed to give a consistently clear, crisp image, and the books finished to the high quality standard for which the Press is recognised around the world. The latest print-on-demand technology ensures that the books will remain available indefinitely, and that orders for single or multiple copies can quickly be supplied.

The Cambridge Library Collection brings back to life books of enduring scholarly value (including out-of-copyright works originally issued by other publishers) across a wide range of disciplines in the humanities and social sciences and in science and technology.

A Monograph of the British Carboniferous Trilobites

Henry Woodward

Cambridge University Press

CAMBRIDGE
UNIVERSITY PRESS

University Printing House, Cambridge, CB2 8BS, United Kingdom

Cambridge University Press is part of the University of Cambridge.
It furthers the University's mission by disseminating knowledge in the pursuit of education, learning and research at the highest international levels of excellence.

www.cambridge.org
Information on this title: www.cambridge.org/9781108081399

This edition first published 1883–4
This digitally printed version 2015

ISBN 978-1-108-08139-9 Paperback

THE

PALÆONTOGRAPHICAL SOCIETY.

INSTITUTED MDCCCXLVII.

LONDON:

MDCCCLXXXIII—MDCCCLXXXIV.

BRITISH CARBONIFEROUS TRILOBITES.

DIRECTIONS TO THE BINDER.

The Monograph of the British Carboniferous Trilobites will be found in the publications of the Palæontographical Society issued for the years 1883 and 1884.

Cancel the title-pages of the separate parts in the volumes for the years 1883 and 1884 (dated 1883 and 1884) and *substitute* the general title-page (dated 1883—1884) now provided, and follow the order of binding given in the accompanying table of pages, plates, and dates.

ORDER OF BINDING AND DATES OF PUBLICATION.

	PAGES	PLATES	ISSUED IN VOL. FOR YEAR	PUBLISHED
Part II	Title-page "1883—1884"			
Part I	1—38	I—VI	1883	October, 1883
Part II	39—86	VII—X	1884	December, 1884

A MONOGRAPH

OF THE

BRITISH CARBONIFEROUS

TRILOBITES.

BY

HENRY WOODWARD, LL.D., F.R.S., F.G.S., F.Z.S., F.R.M.S.,

KEEPER OF THE DEPARTMENT OF GEOLOGY IN THE BRITISH MUSEUM (NATURAL HISTORY);
VICE-PRESIDENT OF THE PALÆONTOGRAPHICAL SOCIETY;
MEMBER OF THE LYCEUM OF NATURAL HISTORY, NEW YORK; AND OF THE AMERICAN PHILOSOPHICAL SOCIETY, PHILADELPHIA; HONORARY MEMBER OF THE YORKSHIRE PHILOSOPHICAL SOCIETY; OF THE GEOLOGISTS' ASSOCIATION, LONDON; OF THE GEOLOGICAL SOCIETIES OF EDINBURGH, GLASGOW, AND NORWICH; CORRESPONDING MEMBER OF THE GEOLOGICAL SOCIETY OF BELGIUM; OF THE IMPERIAL SOCIETY OF NATURAL HISTORY OF MOSCOW; OF THE NATURAL HISTORY SOCIETY OF MONTREAL; AND OF THE MALACOLOGICAL SOCIETY OF BELGIUM.

LONDON:

PRINTED FOR THE PALÆONTOGRAPHICAL SOCIETY

1883—1884.

PRINTED BY
J. E. ADLARD, BARTHOLOMEW CLOSE.

PRELIMINARY NOTICE.

The late Mr. J. W. Salter, in Part I of his 'Monograph of British Trilobites' (p. 2, Pal. Soc., 1864), gave a preliminary classification of Trilobites, in which he placed under the Proetidæ the following genera:—Phillipsia, Griffithides, Brachymetopus, Proetus, and Phaeton[1], and in a "Notice to Correspondents" (facing p. 177 of his last Part, IV, 1867) he expressed an opinion that "it would be in some respects advisable to go on (in Part V) with the highest and most compact group of the smooth-eyed Trilobites, viz. the Proetidæ." Unhappily his labours were brought to a close by death, and I have taken up the unfinished narrative.

After much hesitation, it seemed most convenient to commence with the Trilobites of the Carboniferous Limestone, because, with the exception of the genus *Proetus*, they form a group by themselves which are only to be met with in one formation and in a limited area, at least in this country; and they also very greatly needed to be re-examined and carefully figured.

HENRY WOODWARD.

British Museum (Natural History),
Cromwell Road, London, S.W.;
August 20th, 1883.

[1] Made a synonym of the genus *Proetus*, by its author Barrande.

THE TRILOBITES

OF THE

CARBONIFEROUS LIMESTONE.

INTRODUCTION.

Considerable confusion has hitherto existed in the nomenclature of the various species of Trilobites from the Carboniferous Limestone series, partly arising from the fact of the very near affinities actually existing between the genera of *Phillipsia* and *Griffithides*, partly from the too often fragmentary condition of the specimens obtained, but also largely due to the unsatisfactory figures which accompany the descriptions of most of the early writers on the fossils of this group.

Thanks to the labours of Portlock, M'Coy, Valerian von Möller, Traquair, and others, many of the difficulties to a classification of these, our latest Trilobites, have been removed, and much help afforded in the task of unravelling the tangled skein of synonymy woven by Russian, German, French, Belgian, English, and Irish palæontologists during the past sixty years.

I have given, in the following pages, in chronological order, short notices of the principal works in which the species of Carboniferous Trilobites have been referred to, and in the subsequent descriptions I have attempted, to the best of my ability, to affix to each its proper generic and specific place.

I should but ill fulfil my duty were I to omit to return thanks to the many friends who have assisted me with the loan of specimens for this Monograph; notably I would mention Prof. T. McKenny Hughes, M.A., F.G.S., Woodwardian Professor of Geology in the University of Cambridge, who has most liberally placed the whole of the extensive series of Carboniferous Trilobites belonging to the Woodwardian Museum in my hands for examination and figuring.

I am equally indebted to Prof. A. Giekie, LL.D., F.R.S., F.G.S., Director-General of the Geological Survey of Great Britain and Ireland, and Prof. Edward

Hull, M.A., LL.D., F.R.S., Director of the Irish Branch, for permitting me the same opportunities of studying and figuring the beautiful type-series of forms out of the Jermyn Street Museum and the Museum of the Geological Survey in Dublin. To Dr. R. H. Traquair, F.R.S., the Rev. E. O. de la Hey, John Aitken, Esq., Joseph Wright, Esq., and my kind friends in Glasgow, Messrs. John Young, James Thomson, Robert Craig, J. Smith, and others, who have so willingly entrusted me with the choice specimens from their private collections for my work, I am especially thankful.

In a table further on, I give a list of the Divisions of the Carboniferous series of deposits as recognised in England, Scotland, and Ireland, in compiling and preparing which I am indebted to the kindness of my colleagues, the Messrs. R. Etheridge (father and son); also to Mr. Robert Etheridge, jun., for much valuable assistance in preparing the accompanying Bibliography.

BIBLIOGRAPHY

OF THE

TRILOBITES

OCCURRING IN THE

CARBONIFEROUS LIMESTONE, &c.

---

1. 1809. Mr. W. MARTIN, in his 'Petrificata Derbiensia,' publishes, under the name of *Entomolithus* (*Oniscites*) *Derbiensis*, the earliest description extant of a Carboniferous Trilobite. Under the genus *Phillipsia*, we give Martin's careful and interesting description of this fossil, which he considered to be "an insect, related to *Oniscus*."

2. 1822. MM. ALEX. BRONGNIART and A. G. DESMAREST, in their 'Histoire Naturelle des Crustacés Fossiles,' p. 145, pl. iv, fig. 12, under the name of *Asaphus*, figure a pygidium of *Phillipsia* from the black Carboniferous Limestone in the environs of Dublin.

3. 1823. Baron ERNST F. VON SCHLOTHEIM, in his 'Nachträge zur Petrefactenkunde,' Gotha (ii Abtheil., pp. 42–3), refers to a pygidium of a Trilobite which he also figures on plate xxii, fig. 6, under the name *Trilobites*, *Asaphus pustulatus* (vel *pustulosus*, see explanation of plate accompanying Atlas, p. 22), said to be from the youngest Upper Transitional Limestone of the Eifel. This is no doubt a Devonian Trilobite, but the name having been adopted by de Koninck it has become incorporated in the nomenclature of these later species with which it has no affinities.

4. 1825. EDOUARD D'EICHWALD, in his 'Geognostico-zoologicæ per Ingriam marisque Baltici Provincias, nec non de Trilobitis observationes' (Casani), notices and figures, at p. 54, tab. iv, figs. 4 and 5, two pygidia of Trilobites, which he names, "*Asaphus Brongniarty*" (*sic*), fig. 5, and "*A. Eichwaldi*," fig. 4, on the authority of Fischer de Waldheim.

5. 1829. FRIEDERICH HOLL, in his 'Handbuch der Petrefactenkunde' (Dresden), p. 176, records *Asaphus* (*Phillipsia*) *Brongniarti*, and *A.* (*Phil.*) *Eichwaldi*, Fischer, in his list of Trilobites.

6. 1832. A. H. DUMONT, in his 'Mémoire sur la Constitution Géologique de la Province de Liége,' 'Mémoires Couronnés par l'Académie Royale de Bruxelles,'

tome viii, p. 353, gives in his table of fossils, *Calymene Tristani*, and *C. macrophthalmus* (which probably are equivalent to *G. globiceps* and *P. Derbiensis*) from the Upper Limestone of Richelle, Belgium.

7. 1836. Prof. JOHN PHILLIPS, in his 'Illustrations of the Geology of Yorkshire,' part ii, pp. 239, 240, pl. xxii, figs. 1—20, notices and names eight species of Carboniferous Trilobites from Yorkshire, Derbyshire, and Ireland. The originals of Phillips' species are very fragmentary and the figures are not good.

8. 1836. The Rev. Prof. BUCKLAND, in his 'Bridgewater Treatise,' vol. ii, p. 74, pl. 46, figs. 10 and 11, notices and figures two pygidia of Trilobites from the Carboniferous Limestone of Dublin and Northumberland which he names *Asaphus gemmuliferus*, and *A. caudatus* respectively.

9. 1830-37. G. FISCHER DE WALDHEIM, in his 'Oryctographie du Gouvernment de Moscou' (Moscow), p. 121, pl. xii, figs. 1 and 2, reproduces the two pygidia figured by Eichwald, under the name of *A. Eichwaldi*, and states that he considers them "as one and the same species, for which he retains the name of Eichwald; the more so as another Trilobite already bears the name of Brongniart."

10. 1838. Prof. HENRI MILNE-EDWARDS, in the 2nd edition of Lamarck's Histoire Naturelle des animaux sans vertèbres, présentant les caractères,' &c., vol. v, p. 234, quotes *Asaphus globiceps* (= *Griffithides globiceps*) and *P. seminifera*, from the Carboniferous Limestone.

11. 1839. DR. H. F. EMMRICH, 'de Trilobitis, Dissertatio petrefactologica,' etc. (Berlin), notices several Carboniferous Limestone Trilobites, viz., *Asaphus Dalmani* (= *Phillipsia Derbiensis*); *As. globiceps* and *Calymene* (= *Griffithides globiceps*).

12. 1840. EDOUARD d'EICHWALD, published in the 'Bulletin scientifique publié par l'Academie Impériale des Sciences de St. Pétersbourg;' a paper entitled, "die Thier- und Pflanzenreste des alten rothen Sandsteins und Bergkalks im Nowogordischen Gouvernment," in which he notices (p. 4) a Carboniferous-Limestone Trilobite from Bystriza, under the name of *Otarion Eichwaldii*, Fischer, but without a figure (referred to *Ph. mucronata* by von Möller).

13. 1841. PROF. L. G. DE KONINCK, in the 'Nouveaux Mémoires de l'Académie Royale de Sciences de Bruxelles' (tome xiv, pp. 1—20, with plate), publishes two species of Trilobites from the Carboniferous Limestone, namely, *Asaphus gemmuliferus* and *A. Brongniarti.*

14. 1843. CAPT. (*afterwards* GENERAL) J. E. PORTLOCK, R.E., F.R.S., in his 'Report on the Geology of the county of Londonderry and parts of Tyrone and Fermanagh' (Dublin), pp. 305—313, pl. xi and xxiv, gives careful descriptions of eight species of *Phillipsia* and *Griffithides*, with figures of the same.

15. 1843. DR. HERMANN BURMEISTER, in 'Die Organisation der Trilobiten

aus ihren lebenden Verwandten entwickelt,' etc. (Berlin), pp. 117—139, describes *Æonia*, sp. (= *Phillipsia*); *Archægonus* (= *Griffithides*) *globiceps*; and *Asaphus* (= *Phillipsia*) *pustulatus*.

16. 1843. PROF. DR. GOLDFUSS, in 'Leonhard und Bronn's Neues Jahrbuch,' &c., 8vo, pp. 537—567, gives his "Systematische Uebersicht der Trilobiten," etc. pl. iv, v, vi, and notices *Asaphus Dalmani* (= *Ph. Derbiensis*), *Gerastos Brongniarti* (= *Ph. Eichwaldi*), *Phacops* (= *Griffithides*) *globiceps*, etc., pp. 558—565.

17. 1843. FR. WILH. HÖNINGHAUS, in his 'Trilobites der geognostischen Sammlung' (Crefeld), p. 7, cites *Asaphus Dalmani* (= *Phillipsia Derbiensis*).

18. 1843. J. B. J. D'OMALIUS D'HALLOY, in his 'Précis élémentaire de Géologie' (Paris), p. 515, cites *Asaphus Brongniarti* (= *Ph. Eichwaldi*).

19. 1844. PROF. L. G. DE KONINCK, of Liége, in his 'Description des Animaux Fossiles dans le Terrain Carbonifère de Belgique,' 1842-44, pp. 595—607, pl. lii and liii, describes and figures six species of *Phillipsia* and a detached hypostome of the same, from the Carboniferous rocks of Tournay, Visé, etc., Belgium.

20. 1844. PROF. F. M'COY, in a 'Synopsis of the Characters of the Carboniferous Limestone Fossils of Ireland' (Dublin), pp. 160—163 (pl. iv), describes six species of *Griffithides* and ten species of *Phillipsia*, but *without* localities.

21. 1845. MURCHISON DE VERNEUIL and DE KEYSERLING, in their 'Géologie de la Russie d'Europe,' etc., vol. ii, 4to (Paléontologie by M. de Verneuil), pp. 376—378, describe a pygidium of *Ph. Eichwaldi*, from the Valdai, and another pygidium named *Ph. Ouralica*, from the Ural Mountains.

22. 1845. DR. H. F. EMMRICH, in 'Leonhard und Bronn's Neues Jahrbuch für Mineralogie, Geognosie, Geologie und Petrefaktenkunde' (pp. 18—62), "Ueber die Trilobiten" (with a plate), notices several Carboniferous species; notably *Phillipsia æqualis*, *Ph. ornata*, and *Griffithides globiceps*.

23. 1846. DR. H. F. EMMRICH's (spelt "Emmerich") paper, "Ueber die Trilobiten," was translated and reproduced by Richard Taylor in his 'Scientific Memoirs,' vol. vi, pp. 253—291.

24. 1846. DR. H. BURMEISTER, 'The Organisation of Trilobites deduced from their living affinities' (edited by Professors Bell and E. Forbes—for the Ray Society, London), is a translation of the work published in Berlin in 1843, but "revised, augmented, and in part re-written by the Author."

25. 1846. ALEX. VON KEYSERLING, in his 'Wissenschaftliche Beobachtungen auf einer Reise in das Petschora-Land im Jahre 1843' (St. Petersburg), quotes "*Phillipsia Eichwaldi* and *P. truncatula*," p. 291, as occurring in the Carboniferous Limestone of Wytschegda and Sopljussa, Petschora.

26. 1846. DR. T. OLDHAM, in the 'Journal of the Geological Society of

Dublin,' vol. iii, part iii, p. 188, pl. ii, figures and describes a very perfect specimen of *Griffithides globiceps* (re-drawn on our Pl. VI, fig. 1 *a*, *b*), from the Carboniferous Limestone of Millicent, Clane, Kildare.

27. 1847. PROF. F. M'COY, in the 'Annals and Magazine of Natural History,' vol. xx, p. 229, founded the genus *Brachymetopus* to contain *Phillipsia* (?) *discors*, M'Coy, *P.* (?) *Maccoyi*, Portlock, and *B. Strzeleckii*, M'Coy, therein described and figured (pl. xii, figs. 1 *a* and *b*).

28. 1848. DR. H. G. BRONN, in the "*Nomenclator*" to his '*Index Palæontologicus*, oder Uebersicht der bis jetzt bekannten fossilen Organismen,' &c. (Stuttgart, 2 vols.), enumerates five species of *Griffithides*, and seventeen species of *Phillipsia*.

29. 1849. DR. H. G. BRONN issued the "*Enumerator*" to his '*Index Palæontologicus*,' giving nine species of *Phillipsia*, and four species of *Griffithides*.

30. 1852. M. JOACHIM BARRANDE, in his great work, 'Système Silurien du Centre de la Bohème,' t. iii, fig. 10, figures *Griffithides* (*Ph.*) *globiceps* from the Carboniferous Limestone.

31. 1852-54. PROF. DR. FERD. ROEMER, in his 'Palæolethæa; Kohlen-Gebirge' ('Bronn's Lethæa Geognostica,' Stuttgart,), gives descriptions and figures (p. 594, taf. ix, figs. 8, 9, 10), *Ph. Derbiensis*, *Grif.* (*Ph.*) *globiceps*, and *Ph. gemmulifera*.

32. 1854. PROF. MORRIS, in the 2nd edition of his 'Catalogue of British Fossils,' enumerates three species of *Brachymetopus*, five species of *Griffithides*, and seven species of *Phillipsia*, from various British and Irish localities.

33. 1855. PROF. F. M'COY, in his 'Systematic description of British Palæozoic Fossils in the Geological Museum, Cambridge' (published in Prof. Sedgwick's 'British Palæozoic Rocks,' Cambridge), describes a species of *Griffithides* (figured in pl. 3 D, figs. 10, 11), and three species of *Phillipsia*, all from British localities.

34. 1855. DR. B. F. SHUMARD, in G. C. Swallow's 'First and Second Annual Reports of the Geological Survey of Missouri,' gives at p. 199 a description of *Phillipsia Meramecensis*, Shumard (pl. B, fig. 9), from the "Archimedes Limestone, Meramec River, Fenton, St. Louis."

35. 1856. MR. GEORGE TATE, of Alnwick, described (in the 'Proceedings of the Berwickshire Naturalists' Club,' p. 234,) a mucronate-tailed Trilobite, from the Carboniferous Limestone of Northumberland, under the name of *G. Farnensis*. (See *Phillipsia Eichwaldi*, var. *mucronata*.)

36. 1858. DR. B. F. SHUMARD and G. C. SWALLOW, in their "Descriptions of New Fossils from the Coal-measures of Missouri and Kansas," in 'Transactions of the Academy of Sciences, St. Louis, Missouri,' vol. i, No. 2, record *Phillipsia Mis-*

souriensis, *Ph. major*, and *Ph. Cliftonensis*, said to be from the "Upper and Middle Coal-Measures." They are not accompanied by figures.

37. 1858. Dr. GEO. G. SHUMARD, in his "Observations on the Geological Formations in New Mexico" ('Trans. Acad. Sci. of St. Louis'), pp. 273—297, describes (at p. 296) a new species of *Phillipsia* under the name of *Ph. perannulata*, from the White Limestone, Guadaloupe Mountains, New Mexico.

38. 1860. M. VON GRÜNEWALDT, in his 'Beiträge zur Kenntniss der sedim. Gebirgsformation des Urals' (St. Petersburg), notices two species of Carboniferous-Limestone Trilobites, from Russia, viz., *Ph. Derbiensis* (p. 139, t. v, fig. 12) and *Ph. indeterminata* (p. 140, t. v, fig. 10).

39. 1860. M. EDOUARD D'EICHWALD, in his 'Lethæa Rossica ou Paléontologie de la Russie' (vol. i, pp. 1435—1441, and Atlas of plates, tab. liv, figs. 8—12), gives descriptions of eight species of Carboniferous Trilobites, all of which he refers to the genus *Griffithides;* but his figures are most unsatisfactory and cannot be relied upon at all.

40. 1862. Prof. M'COY's 'Synopsis of the Characters of the Carboniferous-Limestone Fossils of Ireland' was *re-issued* at this date, by Sir RICHARD GRIFFITH, with a new Title-page, and an Appendix, of the Localities of the Irish Carboniferous-Limestone Fossils, pp. 209—271.

41. 1863. Prof. Dr. FERD. ROEMER, in the 'Zeitschrift der Deutsch. geologischen Gesellschaft' (p. 570, t. xiv, fig. 1 *a*, *b*), describes and figures *Phillipsia mesotuberculata*, from the Carboniferous formation of Königs-Grube, Silesia.

42. 1863. Mr. E. BILLINGS, in the 'Canadian Naturalist and Geologist' (vol. viii, p. 209), gives a "Description of a new species of *Phillipsia* from the Lower Carboniferous Rocks of Nova Scotia," under the name of *Ph. Howi.*

43. 1863. Mr. ALEX. WINCHELL, in the 'Proceedings of the Acad of Nat. Sciences, Philadelphia' (part. 1 p. 24), describes two species of *Phillipsia*, viz., *Ph. insignis*, Winch., and *Ph. Maramecensis ?* Shum., from the Burlington Limestone, Burlington, Iowa.

44. 1865. Messrs. MEEK and WORTHEN (in the 'Proceeds. of the Acad. Nat. Sci. Philadelphia') describe at p. 268, three species of Carboniferous Trilobites, viz.:—*Phillipsia* (*Griffithides*) *Portlocki*, *Ph.* (*G.*) *scitula*, and *Ph.* (*G ?*) *Sangamonensis* from Illinois, the two last-named from the Upper part of the Coal-Measures, Springfield, Ill.

45. 1865. Mr. ALEX. WINCHELL (in the 'Proc. Acad. Nat. Sci. Philadelphia,' p. 132) describes two species of *Phillipsia*, viz., *Ph. doris*, Win., and *Ph. Rockfordensis*, Win., from the Goniatite-limestone of Indiana.

46. 1865. Messrs. J. W. SALTER and HENRY WOODWARD prepared a 'Chart of Fossil Crustacea,' the figures engraved by J. W. Lowry, and accompanied by 'a Descriptive Catalogue of all the genera and species figured' (nearly 500 in number).

Figures and references are given to twelve species of Carboniferous Trilobites, mostly British.

47. 1866. Dr. H. B. Geinitz, in his 'Carbonformation und Dyas in Nebraska' (Dresden, pp. 102, with five plates), figures and describes a *Phillipsia*, from the limestone of Plattesmouth, Nebraska (p. 1, pl. i, fig. 1).

48. 1867. Valerian von Möller contributed to the 'Bulletin de la Société Impériale des Naturalistes de Moscou,' a paper, "Ueber die Trilobiten der Steinkohlenformation des Urals," in which he figures and describes seven species of Carboniferous Trilobites from Russia, and gives a most excellent catalogue of all the species already described from that formation.

49. 1867. Sir R. I. Murchison, in his 'Siluria' (4th edition, p. 209, Fossils (79), figs. 1 and 2), gives a woodcut of a head of *Brachymetopus Ouralicus*, de Vern., and of an entire specimen of *Phillipsia pustulata*, the figures are too small, however, to show the characters clearly. This also appeared in the 3rd edition, p. 283 (1854).

50. 1868. Mr. C. Fred. Hartt, A.M., in the 'Canadian Naturalist and Geologist' (Montreal, New Series, vol. iii), notices at p. 214 *Phillipsia Vindobonensis* from the Carboniferous Limestone of Nova Scotia.

51. 1869. Prof. R. H. Traquair, M.D., F.R.S., in the 'Journal of the Royal Geological Society of Ireland,' Dublin, contributes a most valuable paper on *Griffithides* (*Phillipsia*) *mucronatus* with excellent figures of the same.

52. 1870. Messrs. Meek and Worthen published (in the 'Proceed. Acad. Nat. Sciences, Philadelphia') a paper with "Descriptions of New Species and Genera of Fossils from the Palæozoic Rocks of the Western States," in which, at p. 52 they describe two new species of Trilobites, *Phillipsia tuberculata*, M. and W., and *Ph.* (*Griffithides*) *bufo*, M. and W., the former from the Lower Carboniferous of Illinois, and the latter from Indiana.

53. 1872. F. B. Meek, in his 'Report on the Palæontology of Eastern Nebraska' (at p. 238, plate iii, fig. 2 *a*, *b*, *c*), figures and describes *Phillipsia major*, Shumard, from the Upper Coal-Measures of Clinton County, Missouri, and on Vermilion River, Kansas.

54. 1872. Mr. R. Etheridge, F.R.S., in an Appendix to R. Daintree's paper "On the Geology of Queensland" ('Quarterly Journal of the Geological Society of London,' 1872, vol. xxviii, p. 338, pl. xviii, fig. 7), describes and figures a Carboniferous Trilobite under the name of *Griffithides dubius*, Eth. The specimen is obscure.

55. 1873. Messrs. F. B. Meek and A. H. Worthen, in their 'Palæontology of Illinois' (being vol. v of the 'Geological Survey of Illinois,' A. H. Worthen, Director; Springfield, Illinois, pp. 525—529, and pp. 612—618), describe and figure *Proetus* (*Phillipsia*) *ellipticus* (pl. 14, fig. 8), *Phillipsia* (*Griff.*) *Port-*

lockii, *Ph.* (*Grif.*) *bufo*, *Ph.* (*Grif.*) *scitula*, and *Ph.* (*Grif.*) *Sangamonensis* (pls. 19 and 32), the two last-named being from the Upper Coal-Measures of Illinois.

56. 1874. Dr. H. TRAUTSCHOLD (in 'Nouv. Mémoires de la Soc. Imp. des Naturalistes de Moscou,' tome xiii, p. 300) describes three species of *Phillipsia*, viz. *Ph. globiceps*, *Ph. Grunewaldti*, and *Ph. pustulata*, Schlot., from the Carboniferous Limestone of Mjatschkowa, Russia.

57. 1875. F. B. MEEK, Palæontologist, in 'Report of the Geological Survey of Ohio' (vol. ii, part ii, Palæontology, p. 323, plate 18, fig. 3), describes *Phillipsia* (*Griffithides ?*) *Lodiensis*, Meek—probably a *Brachymetopus*.

58. 1875. Mr. ROBERT ETHERIDGE, F.R.S., in a Table of Fossils accompanying "The Geology of the Burnley Coalfield" ('Memoirs of the Geological Survey, England and Wales,' p. 178), records one species of *Griffithides* and five species of *Phillipsia* from the Carboniferous Limestone of Bolland and Clitheroe.

59. 1875. Mr. W. H. BAILY, in his 'Figures of Characteristic British Fossils with Descriptive Remarks' (London, vol. i, pp. lxxiii and lxxiv, and p. 118, pl. 41, figs. 1, 2, and 3), figures and notices *Brachymetopus Ouralicus*, *Phillipsia pustulata*, and *Griffithides globiceps*, from several Irish and English localities.

60. 1875. Prof. Dr. KÖRBER (in the 'Sitzungsberichte der math.-naturw. Classe der Kaiserl. Akad. der Wiss. Wien,' Band lxxi, Abth. i, at p. 529, taf. 1, fig. 1) notices *Phillipsia Grunewaldti*, Möll., from the Coal-formation of Barents Island, Nova-Zembla.

61. 1876. M. FALY, at a meeting of the Société Géologique de Belgique (19th March), exhibited a specimen of a Trilobite from the phtanite coal of Casteau, near Mons, closely related to *Phillipsia globiceps*, Phillips.

62. 1876-77. PROF. L. G. DE KONINCK, in his 'Recherches sur les Fossiles Paléozoïques de la Nouvelle-Galles du sud (Australie);' (Bruxelles, pp. 348—353), notices and figures *Phillipsia seminifera*, Phil., *Griffithides Eichwaldi*, G. Fischer, and *Brachymetopus Strzeleckii*, M'Coy.

63. 1877. DR. H. WOODWARD, F.R.S., published 'A Catalogue of British Fossil Crustacea, with their Synonyms and the Range in time of each genus and order' (printed for the Trustees of the British Museum, London, pp. xii and 156), containing three species of *Brachymetopus*, four of *Griffithides*, and eight of *Phillipsia*, found in the Carboniferous Limestone of Great Britain and Ireland.

64. 1878. MR. ROBERT ETHERIDGE, JUNR., in his 'Catalogue of Australian Fossils' (Cambridge), p. 232, notices the occurrence of two species of *Griffithides* (*G. dubius* and *G. Eichwaldii*), and two of *Phillipsia* (*P. parvula* and *P. seminifera*, p. 42), from Queensland and New South Wales.

65. 1878. DR. J. W. DAWSON, F.R.S., in his 'Acadian Geology' (London, pp. 824), at p. 313 gives a woodcut and description of *Phill. Howi*, and a description of *Phillipsia Vindobonensis*, Hartt.

66. 1879. PROF J. D. DANA, in his 'Manual of Geology' (3rd edition, New York), p. 304 and p. 308, refers to the Carboniferous Trilobites *Phillipsia, Griffithides*, and *Brachymetopus*, and at p. 342 he records *Ph. Missouriensis, P. major*, and *P. Cliftonensis*, Shumard, from the Upper Coal-measures of Missouri; and *P. scitula*, M. and W., common in Illinois and Indiana. At p. 308 he gives a figure of *Phil.* (*Griffithides*) *seminifera* (after De Koninck), and quotes also *P. pustulata*, as occurring in the Irish Rocks.

67. 1879. PROF. H. A. NICHOLSON, in his 'Manual of Palæontology' (2nd edition), 2 vols., p. 371, gives the distinguishing characters of *Phillipsia* and *Griffithides*, but figures *Phillipsia* (*Griffithides*) *seminifera* (reproduced from Dana's 'Geology.')

68. 1879. PROF. A. VON KOENEN, of Marburg, in the 'Neues Jahrbuch für Mineralogie, Geologie und Palaeontologie' (p. 309), gives a description of "The Culm Fauna from Herborn," consisting of more than fifty species; and he describes two species of *Phillipsia*, viz. *Ph. æqualis*, H. von Meyer, sp., and *P. latispinosa*, Sandberger, sp.

69. 1879. MR. ROBERT ETHERIDGE, JR., in 'Explanation of Sheet 31, Memoirs of the Geological Survey of Scotland' (p. 81), refers to the occurrence, in "The shale above the Castlecary Limestone," of remains of a species of *Phillipsia*, nearly allied to *P. pustulata*, Schlot., or *P. seminifera*, Phillips.

70. 1882. PROF. A. GEIKIE, LL.D., F.R.S., in his 'Text-book of Geology' at p. 724, under the Carboniferous Fauna, observes, "Trilobites now almost wholly disappear, only two or three genera of small forms (*Griffithides, Phillipsia, Brachymetopus*) being left."

Family PROETIDÆ.

This family comprises the one Silurian genus Proetus,[1] and the three Carboniferous genera—*Phillipsia*, *Griffithides*, *Brachymetopus*, which form the subject of our present memoir.

Genus I.—Phillipsia, *Portlock*, 1843.

General form oval; glabella with nearly parallel sides, marked by either two or three short lateral furrows; the posterior angles, forming the basal lobes, always separated by a circular furrow from the rest of the glabella; eyes large, reniform, surface delicately faceted;[2] cervical furrow deep; free cheek separated from the glabella by the axal suture which forms an acute angle with the circular border of the cheek in front of the glabella; whilst the facial suture cuts obliquely across the posterior margin, just behind the eye, leaving a small pointed portion fixed to the glabella by the neck-lobe; angles of cheeks more or less produced, margin of head incurved forming a striated and punctated rim. Thoracic segments nine in number, the axis distinctly marked off from the side-lobes or pleuræ by the axal furrows; the abdomen, or pygidium, usually with a rounded border, the axis composed of from 12 to 18 coalesced segments.

The following is General Portlock's description of the genus *Phillipsia*, taken from his 'Report on the Geology of Londonderry, &c.' (8vo. 1843, pp. 305, 306).

" General form, oval. Cephalo-thorax divided into three compartments by the elevation of the glabella above the plane of the cheeks. Glabella, bounded on the sides by nearly parallel lines, and rounded in front; the general form approaching

[1] M. Barrande suggests that Dr. Sandberger's genera *Trigonaspis* and *Cylindraspis*, from the Devonian of Nassau, also belong to the genus *Proetus*. Prof. Dr. Ferd. Roemer has figured and described four species of *Proetus* from the Harz, and Messrs. Meek and Worthen have named a species *Proetus* from the Lower-Carboniferous series of Jersey Co. Illinois, so that *Proetus* serves to connect this otherwise detached group of Mountain-Limestone forms with their relatives in the Silurian and Devonian formations.

[2] Messrs. Meek and Worthen, in their description of two Carboniferous Trilobites from Illinois and Indiana, remark, "eyes apparently smooth, but showing, when the outer crust is removed, numerous very minute lenses beneath," 'Geol. Surv. Illinois,' vol. v, "Palæontology," 1873, 4to, pp. 528, 529. This observation may serve to explain the fact that many specimens do not show the faceted surface at all clearly; this is especially the case in the genus *Griffithides*. Emmrich believed it possible to use this character of the external surface of the eyes of Trilobites as a means of classification, but I have not been able to accept his proposed arrangement based on this structure. (See Emmrich, *op. cit.*, 1845, in "Bibliography," p. 5.)

to cylindrical; marked on the sides by three sets of segmental lines, corresponding to and representing the cephalo-thoracic furrows of *Phacops* and *Calymene*, but which are not present in *Proetus*. The first, from the base, of these lines is rounded, so as to include a circular space corresponding to the tubercle of the lower cephalo-thoracic segment of *Calymene;* the second is also slightly curved; the third nearly straight. Between the two sets of lines there is a space equal to more than one-third of the whole breadth of the glabella, corresponding to the similar axal or connecting space in the glabella of *Phacops* and *Calymene*. The cheeks, slightly convex; in form spherical triangles. Eyes, lunate; situated on the cheeks, but very near to the glabella, to the axis of which their chord or longer axis is parallel; surface under a magnifier very finely reticulated. Neck-furrow deep. Wings, or margins, separated from the front and cheeks by an imperfect line or furrow which is connected with the neck-furrow, and end posteriorly in sharp angles, more or less prolonged as spines. The margin is turned down, and forms, on the under surface, a narrow, striated rim.

"Thorax composed of nine segments, or, including the neck-segment, ten; and it requires some care to avoid mistaking the first pygidial for the last thoracic segment as it is constructed on the same principle as those of the thorax, though its parts are fused together and form one whole with the other pygidial segments. The pleuripedes are compound as in *Phacops*, *Calymene*, *&c.*

"Pygidium exhibiting, both in the axal and lateral lobes, distinct segments by which it is further separated from *Proetus*. The axal lobe and the lateral segments do not extend to the external edge, but have a considerable marginal space. It may be here remarked that as this form of pygidium recurs in another Mountain Limestone genus, although the corresponding forms of cephalo-thorax are strikingly different, the genus cannot be satisfactorily determined from a specimen of the pygidium alone."

1. Phillipsia Derbiensis, *Martin*, sp., 1809. Plate I, figs. 1—9.

Entomolithus (Oniscites) Derbiensis, *Martin*. Petrif. Derb., t. xlv, figs. 1 and 2, 1809.

Asaphus raniceps, *Phillips*. Geol. Yorks., vol. ii, t. xxii, figs. 14 and 15, 1836.

Phillipsia Derbiensis, *De Koninck*. Anim. Foss., t. liii, fig. 2, 1842.

— Jonesii, *var.* seminifera, *Portlock*. Rep. Geol. Londonderry, &c., p. 308, t. xi, fig. 5, 1843.

— Derbiensis, *Morris*. Cat. Brit. Foss., p. 114, 1854.

— Jonesii, *var.* seminifera, *M'Coy*. Brit. Pal. Foss., p. 183, 1855.

PHILLIPSIA DERBIENSIS, *Salter & Woodw.* Cat. and Chart Foss. Crust., p. 16, fig. 111, 1865.
— — *H. Woodw.* Cat. Brit. Foss. Crust., p. 55, 1877.

Glabella smooth, somowhat gibbous, in front; sides nearly straight, with two short furrows near the front of the eye, and a circular furrow around the basal lobe at each posterior angle of the glabella; neck-furrow deep, neck-lobe rather broad, with one small tubercle on centre; fixed cheek very small; facial suture oblique, leaving a small angular portion attached to the neck-lobe on either side. Eyes very large in proportion to head; reniform, smooth, but when well preserved showing a fine and minutely-faceted surface. Facial suture uniting with outer border of free cheek, and forming a very acute angle, where it joins the glabella in front and a less acute angle behind the eye, where it unites with the posterior border. A broad groove or furrow surrounds the free cheek running exactly parallel to its own outer border; the posterior angles of the head project slightly backwards, but are not produced into cheek-spines. The incurved under margin of the shield is finely striated as well as punctated. The hypostome (seen *in situ* in one of our specimens, Pl. I, figs. 4 *a*, *b*) is large; the mesial lobe is broad and spatulate, the surface being finely striated with wavy longitudinal lines; the lateral lobes or alæ are small, smooth, and pointed.

The thorax, which is roundly arched, consists of nine smooth and well-defined segments, the first only having a minute tubercle on the centre. The axis of the thorax, which next the head is considerably broader than its side lobes, diminishes gradually in breadth backwards to the pygidium, where it is only equal to its pleuræ in breadth; the pleuræ, which are smooth, are all faceted to enable the animal to roll itself up into a ball. The axis of the abdomen, or pygidium, shows it to be composed of thirteen coalesced segments, the pleuræ being united in a rounded shield, the border of which is smooth, as the ribs die out before they quite reach the margin. There is a faint tendency to ornamentation on the axis of the tail.

Formation.—In Carboniferous Limestone, and in "Rotten-stone" band.

Localities.—Bolland and Settle, Yorkshire; Castleton, Derbyshire; "Rotten Stone," Matlock, Derbyshire; Longnor; Arnside; Blackrock and Little Island, Co. Cork; Middleton; Carnteel, Tyrone; Castlepollard, West Meath, and Limerick, West of Dromore Wood.

This is undoubtedly the earliest species of Carboniferous Trilobite recorded, and is probably next in historical antiquity to the famous "Dudley Locust," *Calymene Blumenbachii*, described by Lyttleton in 1750, from the Wenlock (Upper Silurian) Limestone.

Fortunately we are acquainted with several perfect specimens of *P. Derbiensis*, two of which are figured in our Plate I.

Fig. 2 *a*, *b*, is a beautiful and very perfect specimen somewhat curved (as if not quite unrolled) from the Carboniferous Limestone of Longnor, Staffordshire, the original of which is preserved in the Museum of Practical Geology, Jermyn Street.

This specimen, which is imbedded in hard limestone, has all the segments of the body well preserved and united together, a rare occurrence in the Carboniferous Limestone. The entire form is carefully reproduced in the outline, fig. 6.

Two other nearly entire specimens, preserved on a piece of Carboniferous "Rottenstone" from Matlock, Derbyshire, are represented in Plate I, fig. 4 *a*. These are from the National Collection, Natural-History Museum, Cromwell Road. Owing, however, to the soft nature of the matrix no very fine structure can be observed, but in one of the specimens the hypostome (4 *b*) can be distinctly seen *in sitû*.

The specimen represented in Plate I, figs. 1 *a*, *b*, is interesting as the type of Phillips' *Asaphus raniceps*, but I see no reason for separating this little head from *P. Derbiensis*, although, from the fact of its being slightly flattened above, the glabella appears to overhang the border more than in the normal forms.

Two other heads in good preservation, figs. 3 and 5, the former from Settle, Yorkshire (from the Woodwardian Museum, Cambridge), and the latter from Castleton, Derbyshire (from the Cabinet of the Rev. E. O. de la Hey), exemplify well the general character of the head of this species.

Fig. 2 shows the finely faceted character of the eyes, as do also some admirable specimens recently received from Prof. E. Hull, M.A., LL.D., F.R.S., from the Geological Survey Museum, Dublin.

After examination of all the specimens, and a careful consideration of General Portlock's descriptions, I see no reason to retain *P. Jonesii* or *P. Jonesii*, var. *seminifera*, M'Coy, as separate species, distinct from *P. Derbiensis*. In this view, I am happy to be in accord with my friend Prof. Morris, M.A., F.G.S., whose carefully prepared 'Catalogue of British Fossils' is still honoured with a place of respect for the critical care displayed by its author in its compilation, although now nearly thirty years old, and sadly in need of a new edition.

The subjoined is the original description of *Phillipsia Derbiensis*, given by Mr. W. Martin, in his *Petrificata Derbiensia* (1809), and named by him *Entomolithus Oniscites* (*Derbiensis*).[1] "This fossil," he says, "is not frequent in many parts of the country. It is principally met with in the black marble at Ashford, where it very rarely occurs in a perfect state; the head and body being found, for the most part, separate from each other." Martin describes it as "A petrified insect. The original an *Oniscus*. The body oblong-ovate, broad and rounded at the head,

[1] 'Petrif. Derb.,' tab. xlv, figs. 1, 2. In the same work, on plate xlv*, figs. 1, 2, Martin represents what seems to be intended for *Griffithides* (*Phillipsia*) *seminifera*, Phil. sp., hereafter described.

smaller and more pointed at the tail; convex, marginated; the margin entire, or not divided by the segments of the back. Head or thorax large, gibbous, equal in breadth to the abdomen; semicircular in front, with a broad, distinct, striated margin, joining that of the body; behind straight, separated from the back by a transverse line. The surface of the head longitudinally divided into three distinct parts; the middle one of these gibbous, rounded, and, when examined under a glass, apparently somewhat rough or scabrous. Between this part and the back a small protuberance, constantly surmounted by a single minute point or tubercle, which, however, is not visible without the assistance of a magnifier. The lateral portions of the surface of the head are nearly of a triangular form; each furnished near the centre with a large lunated tubercle, discovering, in perfect specimens, a reticulated structure, like that of the eyes of living insects, when magnified. The back of the insect is composed of strong, convex, triarcuate segments, their number varying from twenty to twenty-four, each marked with a line of very minute tubercles. The middle parts of the segments are more elevated than those of each side, and form collectively, down the back, a keel-shaped prominence, which ends somewhat obtusely before it reaches the margin of the tail. The segments of the keel directly transverse, those on each side, particularly near the tail, somewhat oblique. The tail obtuse, entire, and destitute of any appendage.

"The above-described parts are all that are ever present in the fossil.—And as the under side is constantly filled with the stone which constitutes the matrix, it would be impossible to examine the legs and inferior parts of the abdomen, did they remain, which it is evident, however, they do not; the petrifaction being formed merely from the upper shell, or covering of the back and head. . . . .

"The lunated tubercles on the head were apparently invested in the recent subject with a much thinner integument than the other parts of the insect. In perfect specimens the dark colour of the limestone is always seen through the present sparry covering of these protuberances, while the rest of the petrifaction, from the greater thickness of the crust, appears perfectly white and opaque. There can scarcely be a doubt that the parts in question were the eyes in the living animal. Their form, as well as the evident difference of their native covering from that of the body, first led to this conclusion; but what places the matter almost beyond uncertainty is their reticulated structure.—This, with the help of a glass, is sufficiently visible: and we may observe that such a structure, while it proves the nature of the parts where it is found, is also illustrative of the operation under which the mineral change has been effected.—Since only a *slow* and *gradual substitution* of fossil for organic matter could have preserved in the petrifaction a conformation thus minute."

2. PHILLIPSIA COLEI, *M'Coy*, 1844. Plate II, figs. 1—10.

PHILLIPSIA COLEI, *M'Coy*. Synop. Carb. Foss. of Ireland, p. 161, tab. iv, fig. 6, 1844.
— — *Morris*. Cat. Brit. Foss., p. 114, 1854.
— — *H. Woodw.* Cat. Brit. Foss. Crust., p. 55, 1877.

Head-shield broadly-semicircular; glabella but slightly elevated, the central convexity not reaching to the front border, but separated by a broadly-expanded margin which makes the head one third wider in front than at its posterior border; glabella marked by two short lateral furrows and by a small basal lobe on each side, the neck-furrow is rather strongly marked, the neck-lobe is slightly broader than the first free segment; the posterior margin is divided obliquely by the facial suture which runs in a very undulating line between the glabella and the free cheek; eyes large, reniform, no facets visible; cheeks arched, somewhat produced at the posterior angles, surrounded by a furrow parallel to the border; free segments nine; axis very slightly arched, equal to its pleuræ in breadth anteriorly, but diminishing slightly towards the pygidium; pleuræ faceted, extremities slightly produced and recurved: pygidium semicircular, axis slightly arched and composed of twelve coalesced segments; pleuræ only faintly indicated, margin of pygidium smooth and slightly bevelled. Surface of head and body generally (save the extremities of the pleuræ) finely granulated.

A detached hypostome found in the same piece of matrix with one of the Survey specimens has been referred to this species. It is 7 mm. long and 4 mm. broad. It is oblong in form; the alæ are very minute, the central lobe is gibbous and ornamented with five raised concentric striæ or wrinkles, irregularly disposed. (See Pl. II, fig. 6.)

This well-marked species was named by M'Coy after the present Earl of Enniskillen; and, having only been found in Ireland, it has escaped the entanglements of palæontological literature, and is in consequence without synonyms. Although quite distinct from any other species of *Phillipsia*, it is marked by excellent generic characters.

In the peculiar broad, smooth, circular border to the front of the glabella this species approaches nearest to *Ph. truncatula* and *Ph. Eichwaldi*. It differs from *Ph. Derbiensis*, in which the glabella is very gibbous and actually overhangs the front border. But in the broad, short, and flattened form of the pygidium we seem to lose the ordinary tail of the Carboniferous Trilobite and to find a strong resemblance to the pygidium of *Asaphus* and *Ogygia* proper. This leads one to observe that the form of the pygidium appears to be a less constant character and of much less value for classification than the cephalic shield.

The specimens figured in Plate II are all from the Museum of the Geological Survey of Ireland.

Formation.—Carboniferous Limestone.

Localities.—Little Island, Cork; N. E. of Ballintra and Carrickbreeny, Donegal; Doohybeg, Co. Limerick, Ireland.

We subjoin Prof. M'Coy's original description of *Phillipsia Colei,* from his "Synopsis."[1]

"*Specific Characters.*—Elongate, oval; length rather less than twice the width; sides parallel; cephalo-thorax smooth; glabella rounded in front, but not encroaching on the margin, slightly convex, constricted at the sides; cephalo-thoracic furrows distinct, cheeks large, flattened; wings narrow, ending posteriorly in short triangular spines; eyes small, lunate. Thorax: axal lobe rather wider than the lateral ones; each segment having a row of minute, crowded, irregular granulations; pygidium rounded, margin broad, finely granulated, each segment having a row of numerous, crowded, very unequal granulations, larger than those of the thorax.

"This species resembles *P. Kellii,* of Portlock, but is easily distinguished by the character of the granulation of the segments. Length 11 lines, width 6 lines; length of glabella 4 lines, width 6 lines. I have dedicated this elegant fossil to the Earl of Enniskillen."

3. PHILLIPSIA GEMMULIFERA, *Phillips,* sp. 1836. Plate III, figs. 1—8.

ASAPHUS, *sp. indet., Brong. & Desmar.* Hist. Nat. des Crust. Foss., p. 145, pl. iv, fig. 12, 1822.

— "STOKESII," *Fischer.* Oryct. du Gov. de Moscou (footnote, p. 121, *sine descriptione*), 1830–37.

— GEMMULIFERUS, *Phil.* Geol. Yorks., vol. ii, pl. xxii, fig. 11, p. 240, 1836.

— — *Buckland.* Bridgw. Treat., vol. ii, p. 74, pl. 46, fig. 10, 1836.

PHILLIPSIA PUSTULATA, *De Koninck* [*sed non Trilob. pustulatus,* Schlot, 1823]. Descr. Anim. Foss. Terr. Carbonif. de Belg., p. 603, tab. liii, fig. 5, 1842–44.

— KELLII, *Portlock.* Rep. Geol. Lond., p. 307, pl. xi, figs. 1, *a—c,* 1843.

— QUADRISERIALIS, *M'Coy.* Synop. Carb. Foss. Ireland, pl. iv, fig. 8, p. 162, 1844.

— PUSTULATA, *Morris.* (In part only.) Cat. Brit. Foss., p. 114, 1854.

— — *Salter & H. Woodw.* Cat. and Chart Brit. Foss. Crust., p. 55, fig. 109, 1865.

— — *H. Woodw.* Cat. Brit. Foss. Crust., p. 55, 1877.

[1] 'Carb. Foss. Ireland' (1844), p. 161.

General form an elongated oval; head semicircular, glabella rounded anteriorly, the raised central portion marked by two short lateral furrows on each side, and by the usual rounded basal lobes, surrounded in front of the eyes by a rather broad, flat, circular border formed by the fixed cheek, which contracts behind the eyes, but extends obliquely outwards on each side from the neck-lobe, which is rather wider and more strongly marked than the succeeding free segments of the thorax.

Eyes moderately large, reniform, smooth, for, when the faceted surface is visible, the lenses are very minute. Free cheeks terminating in an acute angle on the anterior border, and elongated posteriorly into spines which reach to the fifth thoracic somite. Raised portion of the cheek sparsely granulated, border smooth and broad, under margin striated. The nine free thoracic somites are nearly of equal size, the raised axis being slightly broader than its pleuræ; the axis smooth, pleuræ very minutely granulated and bluntly terminated.

Axis of pygidium composed of sixteen coalesced somites with four to five granules in a row on each axial somite, about thirteen lateral ridges to the pygidium with about six granules on each; margin narrow, plain.

Formation.—Carboniferous Limestone.

Localities.—Bolland, Yorkshire; Clithero, Lancashire; Derbyshire; Kildare; St. Doolagh, Co. Dublin; Limerick, west of Dromore Wood; Hook Head, Wexford; and Little Island, Cork.

It has been a source of no small anxiety to adjust the various claims of authors to priority in the naming of this species.

I was at first led astray by Prof. De Koninck, and, guided by him, had placed *P. gemmulifera*, *P. Kellii*, and *P. quadriserialis* under *P. pustulata*. But upon referring to Schlotheim's "*Trilobites pustulatus*" ('Nachträge zur Petrefact.,' Gotha, 1823), I found not only that his figure could by no possibility be made to accord with any Carboniferous Trilobite, but on referring to the text he states that his specimen was derived "aus dem jüngerem Uebergangs-Kalkstein von der Eiffel;" or from the newer Transition Limestone (Devonian). It is in fact a pygidium of *Phacops*, or *Dalmania*.

Dismissing then *P. pustulata* from the list, it was next necessary to consider the rival merits of *P. gemmulifera* and *P. Kellii* to precedence. "*Asaphus gemmuliferus*" was the name given by Phillips to a pygidium in the "Gilbertson Collection" from Settle, but his representation of it is not satisfactory, and it was with pleasure that I turned to Portlock's figure and able description of *P. Kellii*, based not on a pygidium alone (on which he asserts it is quite unsafe to make a species) but upon an entire specimen. Here I felt at last was solid ground to rest upon. But on looking carefully over the Gilbertson Collection I discovered the original specimen, figured by Phillips as "*A. gemmuliferus.*" Now, this is readily seen, on comparison, to be identical with the pygidium of *Phillipsia Kellii*, and as it does not appear likely to be easily confounded with its nearest relative, *P. truncatula* (which species has 18 tail-segments, whilst *P. gemmulifera* has but 15), according to the laws of priority Phillips' name must stand, although based upon a detached pygidium only.

As far back as 1822 MM. Brongniart and Desmarest, in their 'Histoire Naturelle des Crustacés Fossiles," figured a tail of a Trilobite from the Carboniferous Limestone near Dublin,[1] after a drawing by Mr. Stokes, but they do not name it. Fischer, in his 'Oryctographie du Gouvern. de Moscou,' states[2] that this Trilobite was named by them "*Asaphus Stokesii*," but, although I have made diligent search, I cannot confirm this statement of Fischer's, and it seems certain that Brongniart and Desmarest did not give any specific name or description of this fossil, merely calling it "*Asaphus*." The name *Stokesii* must therefore be attributed to Fischer, and not to Brongniart and Desmarest.

The subjoined is Prof. Phillips' original description of *A.* (*Phillipsia*) *gemmuliferus*, 'Geol. Yorks,' vol. ii, p. 240.

"*Asaphus gemmuliferus*, Phillips.

"Each abdominal lobe ornamented by six longitudinal lines of elevated puncta; the transverse furrows undulate the limb (the cast is nearly smooth). I suppose Brongniart's fig. 12, pl. iv, represents this species."

We extract the following valuable remarks by Portlock on Phillips' *Asaphus gemmuliferus*:

"Professor Phillips ('Geology of Yorkshire') has described eight species of Trilobites, all of which he includes provisionally in the genus *Asaphus*. The genera of two of these species, *Asaphus granuliferus*, pl. xxii, fig. 7, and *Asaphus gemmuliferus*, pl. xxii, fig. 11, cannot be determined with certainty from the figures which represent pygidia only. Of the other species, *Asaphus seminiferus*, pl. xxii, figs. 8, 9, 10, *Asaphus truncatulus*, pl. xxii, figs. 12, 13, and probably *Asaphus obsoletus*, in part, pl. xxii, figs. 3, 5, belong to the present genus. On comparing Phillips' fig. 12 with pl. xi, fig. 1 *b* [of Portlock's Rept. Geol. Lond.], a striking general resemblance will be perceived; he, however, represents the glabella as quadrisulcate, whereas the number of sulci in the Irish specimens is only three, a number consistent with the view here taken of the genus. The posterior angles also of the cephalothorax of Professor Phillips' figure do not project backwards so far as in the Irish species; these two points of difference may, however, be the result of some slight imperfections in Professor

[1] This figure is evidently that reproduced in 'Buckland's Bridgewater Treatise,' vol. ii, p. 74, pl. 46, fig. 10, stated to be from the "Transition Limestone, Dublin," and named by him "*Asaphus gemmuliferus* of Phillips."

[2] In a footnote to p. 121 (op. cit., 1830—37).

Phillips' specimens. The head (fig. 12) does not, of course, belong to the same individual as the abdomen (fig. 13). Between fig. 13 and the magnified representation in fig. 1 *c* a similar resemblance is evident; in fig. 1 *c*, however, the margin (limb of Phillips) is not striated; but it may be presumed, from the remarks already made on the striæ of the pygidia of Trilobites, that there is this difference, simply because in one case the natural cast exhibits the external, in the other the internal surface; as, however, the term *truncatulus* seems to have no peculiar reference to this species, it will be named after the gentleman, Mr. Kelly, who supplied the beautiful specimen figured here" (op. cit., pp. 306, 307).

The beautiful and perfect specimen of *Phillipsia* drawn on Pl. III, fig. 1, is the type of that called *P. Kellii* by Portlock, who thus described it:—[1]

"Elongated oval; length ·9″, breadth ·5″ nearly. *Cephalothorax* elevated, bounded by a flattened margin or wings, which extend at the posterior angles into spines, as far as the fifth thoracic segment. *Glabella* convex, elongated, even in front with the cheeks, but not extending to the edge of the margin; rounded in front, at the sides bounded by lines nearly parallel, yet with a slight curvature inwards. Breadth equal to about one third of the total breadth of cephalothorax; the third or anterior cephalothoracic furrow very faint; *cheeks* large, and the eyes being comparatively small there is a large clear space; *eyes* do not reach the neck furrow behind, nor extend beyond the third furrow in front; neck-furrow deep.

"*Thorax.*—Axal lobe about equal to lateral lobes in breadth; axal segments not marked by tubercles, which are arranged, however, along the arched division of each pleuripes, not along the angular division, and continue up to the knee or bend of the segment, which is strongly marked.

"*Pygidium* rather wider than long, about fifteen axal and thirteen lateral segments, each marked by six small tubercles, so arranged as to form longitudinal lines on the axal and lateral lobes; distinct margin. Cephalothorax, including the neck-segment, one third the whole length; thorax rather less than a third; abdomen rather more" (*op. cit.*, p. 307).

On Pl. III, fig. 8, the artist has endeavoured to represent one of the eyes of a specimen of *Ph. gemmulifera*, obtained by the late Mr. John Rofe, F.G.S., from the Carboniferous Limestone of Clitheroe, Lancashire. The eye is extremely perfect, and measures one millimètre in breadth and three millimètres in length. The surface is beautifully faceted hexagonally, each facet being convex on its exposed surface. Taking the number of transverse facets at 16, and of longitudinal rows at 36, there would be about 576 facets in the eye of this Trilobite.

[1] 'Geology of Londonderry' (1843), p. 307.

4. PHILLIPSIA TRUNCATULA, *Phil.*, sp. 1836. Pl. III, figs. 9—14.

ASAPHUS TRUNCATULUS, *Phillips.* Geol. Yorks., vol. ii, pl. xxii, figs. 12, 13, 1836.
PHILLIPSIA ORNATA, *Portlock.* Rep. Geol. Lond., p. 307, pl. xi, fig. 2 *a*, 1843.
— TRUNCATULA, *M'Coy.* Syn. Carb. Foss. Ireland, p. 163, 1844.

Head-shield broadly arched; glabella twice as long as broad, rounded in front, only slightly elevated; basal lobes rather produced, with three short lateral furrows on each side, two of which are anterior to the compound eyes; neck-lobe distinctly marked, and, like the surface of the glabella, rather closely granulated; fixed cheek narrow behind, forming a small, rounded, palpebral lobe above each eye, and expanding into a wide flat circular border in front of the glabella; eyes reniform, smooth; raised portion of the free cheek sparsely granulated; border smooth and broad, terminating in a strong short cheek-spine which is striated beneath.

Thoracic rings wanting.

Pygidium.—Axis composed of eighteen coalesced somites, with six granulations on each axal segment; coalesced ribs of border also granulated; no distinct border to pygidium.

Formation.—Carboniferous Limestone.

Localities.—Bolland and Settle, Yorkshire; Castleton, Derbyshire; Monaster; Millicent; Limerick; Hook Head, and Malakeede, near Dublin, Ireland.

This species has happily escaped the fate of its predecessor, having only been named twice, by Phillips in 1836, and by Portlock in 1843. The head has one more lateral furrow than the other species of *Phillipsia,* and it has the greatest number of coalesced segments in its pygidium of any Carboniferous form. Its nearest ally is *Ph. gemmulifera,* figured on the same plate (Pl. III).

The following is Prof. Phillips' original description of *Phillipsia truncatula* given in his 'Geology of Yorkshire' (1836, vol. ii, p. 240).

"Depressed, mesial lobe of the head quadrisulcate, bituberculate; the eyes lunate; limb continuous, truncate, with undulating parallel striæ; six lines of elevated puncta on the abdominal lobe."

Portlock says[1] of this species, which he calls *ornata:*

"This specimen is imperfect; the form of the glabella approximates closely to that of the preceding species [*P. Kellii*]; it does not, however, extend, as in it, to the points of the cheeks or posterior edge of the margin, and its furrows are more strongly marked. The furrows are ornamented by tubercles, and the whole surface is granular. The tubercles of the pygidium are more elevated, but its general form is nearly the same as that of *P. Kellii,* and it closely resembles *Asaphus truncatulus.*"

[1] 'Geology of Londonderry' (1843), p. 307.

5. PHILLIPSIA EICHWALDI, *Fischer*, sp. 1825. Pl. IV, figs. 2, 4—11, 13, 14.

ASAPHUS EICHWALDI, *Fischer*. MS. Geognostico-zool. per Ingriam Balt. Prov., p. 54, tab. iv, fig. 4, 1825. (Published by Eichwald.)
— — *Fischer de Waldheim*. Oryctog. du Gouv. de Moscou, p. 121, pl. xii, figs. 1 and 2, 1830–37.
OTARION EICHWALDI, *Eichw*. Bulletin Scient. de St.-Petersbourg, p. 4, "Die Thier. &c., des Bergkalks im Nowogorod. Gouv., &c.," 1840.
PHILLIPSIA CŒLATA, *M'Coy*. Synop. Carb. Foss. Ireland, p. 161, tab. iv, fig. 4, 1844.
— — *Morris*. Cat. Brit. Foss., p. 114, 1854.
GRIFFITHIDES MESO-TUBERCULATUS, *M'Coy*. Brit. Pal. Foss. Cambridge, p. 182, pl. 3D, figs. 10 and 11, 1855.
PHILLIPSIA EICHWALDI, *von Möller*. Trilob. der Steinkohlen. des Ural; Bull. Soc. Imp. des Natur. de Moscou, part 1, p. 121, 1867.
— CŒLATA, *H. Woodw*. Cat. Brit. Foss. Crust., p. 55, 1877.

Head-shield circular, glabella slightly gibbous in front, but not overhanging the fixed border which surrounds its anterior margin, and also forms a rounded palpebral lobe over each eye; basal lobes distinctly marked, rather triangular in form, with two short lateral furrows on each side of the glabella at the back of the eyes. The head is marked by two pores, one on each side of the raised glabella just in front of the eyes (I discuss their nature later on);[1] eyes rather large, and somewhat strongly faceted, the facets being larger than in *Ph. gemmulifera;* neck-lobe broad, marked by a single tubercle on centre, and by a row of fine granulations or minute tubercles along its posterior border, like those on the axis of the thorax, and separated on each side by a strong furrow; raised free cheek, small, but surrounded by a broad, flat margin, bevelled on the edge and striated on the under rim, cheeks produced posteriorly into a long spine, which reaches even to the ninth thoracic somite; the entire head-shield, save the margin, is ornamented by very fine granulations; axis of thorax and abdomen very distinct, each ring being marked by a row of very minute, spine-like granulations along its posterior border; ends of pleuræ roundly terminated. *Pygidium* composed of sixteen coalesced somites, central axis ornamented like thoracic axis; lateral lobes of pygidium about eleven in number; margin smooth; border rounded, and striated with fine parallel lines.

Hypostome.—The hypostome (figs. 4 and 7), which is finely striated, has an oblong, median axis, slightly pointed at the extremity like a heraldic shield. Fig. 7 has lost its alæ, but fig. 4 shows them to have been triangular in form.

[1] See Appendix to this Monograph at end of Part II.

Formation.—Carboniferous Limestone.

Localities.—Bolland, Yorkshire; Derbyshire; Tyrone, Ireland; Lennoxtown, Campsie; Newfield Quarry, High Blantyre, Lanarkshire; Gateside, Beith; Auchenskeith and Bowertrapping, near Dalry; Robroyston, near Glasgow; Auchenbeg, near Lesmahagow; Capelrig, E. Kilbride; Gair, Carluke; Boghead, near Hamilton, Scotland.

6. PHILLIPSIA EICHWALDI, var. MUCRONATA, *M'Coy*, 1844. Pl. IV, figs. 1, 3, 12, and 15.

ASAPHUS CAUDATUS, *Buckland.* Bridg. Treat., vol. ii, p. 74, pl. 46, fig. 11, 1836.
OTARION EICHWALDI, *Eichwald.* Die Thier- und Pflanzenreste des Gouvern. Novgorod, p. 4. (Bullet. scientif. de St.-Petersb.), 1840.
PHILLIPSIA MUCRONATA, *M'Coy.* Synop. Carb. Foss. Ireland, p. 162, tab. 4, fig. 5, 1844.
— EICHWALDI, *Verneuil.* Geol. de Russie, vol. ii, p. 376, tab. xxvii, fig. 14, 1845.
GRIFFITHIDES FARNENSIS, *G. Tate.* Proc. Berwick Nat. Club, p. 234, 1856.
— EICHWALDI, *Eichwald.* Lethæa Rossica, Format. anc., p. 1435; Atlas, pl. liv, fig. 10, 1860.
— (or PHILLIPSIA) EICHWALDI, *Salter & Woodw.* Cat. and Chart Foss. Crust., p. 16, fig. 116, 1865.
PHILLIPSIA MUCRONATA, *Möller.* Ueber die Trilob. der Steinkohl. des Ural, &c. Bull. de la Soc. Imp. des Nat. de Moscou, No. 1, p. 121, 1867.
GRIFFITHIDES MUCRONATUS, *Traquair.* Journ. Roy. Geol. Soc., Ireland, Dublin, pp. 213—218; and Plate, figs. 1—7, 1869.
— — *H. Woodw.* Cat. Brit. Foss. Crust., p. 37, 1877.

In this variety MUCRONATA, the only difference we have been able to detect is in the pygidium, in which the posterior border instead of being rounded, as in *Ph. Eichwaldi*, is produced into a short blunt mucro.

Formation.—Carboniferous Limestone.

Localities.—Beadnell, Northumberland; Settle, Yorkshire; Wilkieston, Fife; Gallowhill, Strathavon; Swindridge and Bowertrapping, near Dalry; Caaf Water, Linn.; Dalry; Garple Water, Muir-kirk; Sculliongour, near Lennoxtown, Campsie; Scotland.

Subjoined is the original description of *Griffithides mucronatus* by Prof. M'Coy, given in 1844.[1]

"*Specific Characters.*—Pygidium semi-elliptical, terminating posteriorly in a

[1] See 'Syn. Carb. Foss. Ireland,' p. 162, pl. iv, fig. 5.

short, mucronate, obtuse point; axal lobe nearly as wide as the lateral ones; about fifteen axal and nine lateral segments; surface smooth.

"This is the only Trilobite I know of in the Mountain-limestone with a mucronate or pointed tail. Length of pygidium four lines, width five lines."

We are indebted to Professor R. H. Traquair, M.D., F.R S., &c., for a very carefully prepared *résumé* of the literature of this species, from which we make the following extract ('Journal of Roy. Geol. Soc. of Ireland,' Vol. II, New Series, p. 213).

"In the year 1844 M'Coy figured, under the name of *Phillipsia mucronata*, the pygidium of a Trilobite from the Irish Carboniferous Limestone, and which he considered as new, remarking that he was acquainted with no other Carboniferous form having the caudal extremity prolonged, as in this instance, into a pointed spine or mucro.[1] But a trilobite, apparently the same as this, had previously been known to Continental observers under a different name; for, in 1840, we find Eichwald[2] mentioning a Carboniferous trilobite from Bystriza, in the Government of Novogorod, in Russia, in which the tail is prolonged into a long pointed spine, and whose cephalic shield has on each side a long process. The name quoted for this species is *Otarion Eichwaldii* (Fischer), but the description is, indeed, sufficiently vague, and is not accompanied by any figure. However, in 1845, Verneuil,[3] in figuring and describing as '*Phillipsia Eichwaldii*' a pygidium, to all appearance belonging to the same species as the Irish specimen, put aside the name '*mucronata*' (M'Coy) as a synonym of '*Eichwaldii*' (Fischer). This example was followed by Bronn[4] and by Morris,[5] and the remains of this species, not uncommon in many British localities, are at present very generally labelled and catalogued as *Phillipsia* or *Griffithides Eichwaldii*.

"On turning back, however, to Fischer's original description and figure of *Asaphus Eichwaldii* from Vereia, in the Government of Moscow, published in 1825,[6] we are surprised to find a *rounded* pygidium represented and described. The head is spoken of as unknown, but as regards the tail we read: 'Cauda depressa, *subrotundata* segmentis tredecim ad quatuordecim, margine angustato, sulco infra profundo.' Again, in 1837,[7] the same pygidium was figured and described by Fischer. The figure looks as if it were taken from the same specimen represented in the former work, although the rounding of the caudal extremity is

[1] 'Synopsis of the Carb. Foss. Irel.,' pl. iv, fig. 5.

[2] "Die Thier- und Pflanzenreste des alten Rothen Sandsteins und Bergkalks im Nowogorodischen Gouvernement." ('Bull. Sci. de St.-Peterb.,' 1840.)

[3] 'Géol. de Russia' (1845), vol. ii, p. 376, t. xxvii, fig. 14.

[4] 'Index Palæontologicus.' Stuttgardt, 1848, vol. ii, p. 958.

[5] 'Cat. Brit. Foss.' London, 1854, p. 109.

[6] Contained in a work by Eichwald, entitled 'Geognostico-zoologicæ, per Ingriam marisque Baltici provincias, nec non de Trilobitis observationes.' Casan, 1825, p. 54, t. iv, fig. 4.

[7] 'Oryctographie du gouvernement de Moscou,' p. 120, pl. 12, fig. 2, *a*, *b*.

not here given so definitely as altogether to preclude us from the idea that a spine might have been there, and had been broken off. What we *read* is, however, definite enough, 'La queue est subconique *arrondie*.'

"Can this, then, be the same species as that in which Eichwald, in 1840, says the tail is prolonged into a long spine? Is *Asaphus Eichwaldii* of Fischer really the *Otarion Eichwaldii* of Eichwald? De Koninck did not think so in 1842, for he gives Fischer's *Asaphus Eichwaldii* as the probable synonym of *Phillipsia* (*Griffithides*) *globiceps* (Phillips sp.), a well-known species with rounded tail.[1] However, before saying anything more in answer to this question, it is necessary to investigate still further the literature of the pointed-tailed species.

"In 1860, Eichwald[2] gave an entire figure of *Griffithides Eichwaldii*, accompanied by a description more in detail than that which he published in 1840. The figure given in the 'Lethæa Rossica' will be seen, however, to present some marked discrepancies with those by which the present paper is illustrated.[3] The cephalic spines are shown of enormous length, the eyes occupy a most remarkable anterior position close to the margin of the cephalic shield, and more than one line or furrow crosses the central part of the head. The number of thoracic rings is nine, of axial segments in the pygidium, eighteen. No reference is made in the description to any ornamentation of the surface.

"In 1867, Valerian von Möller[4] in describing a pygidium of this species, from the neighbourhood of Tschernischkinaja, in the Government of Kaluja in Russia, pointed out the delicate ornamentation of the surface, not before noticed as characteristic of the so-called *Griffithides Eichwaldii*. He criticised Eichwald's figure in the 'Lethæa Rossica' with great severity, going even so far as to doubt its genuineness, and to suspect its being 'a not quite successful restoration in which the characters of two quite different forms occur, that of *Phillipsia mucronata*, and of another hitherto little known species.' To this Trilobite Möller restored M'Coy's name of '*Phillipsia mucronata*,' accepting Fischer's description of '*Asaphus Eichwaldi*' as applying to a round-tailed species. Von Eichwald, in his reply[5] to von Möller's criticism, maintained the identity of Fischer's *Asaphus Eichwaldii* with the mucronate-tailed species, and asserted the presence of a small depression or opening at the end of the pygidium which indicated a broken-off process."

[1] 'Description des Animaux fossiles qui se trouvent dans le terrain carbonifère de Belgique,' p. 600.

[2] 'Lethæa Rossica, Formations anciennes,' p. 1435, Atlas, pl. liv, fig. 10.

[3] Reproduced in our Plate IV, figs. 1 and 3.

[4] "Ueber die Trilobiten der Steinkohlenformation des Ural, nebst einer übersicht und einigen Ergänzungen der bisherigen Beobachtungen über die Kohlentrilobiten im Algemeinen." Von Valerian von Möller. 'Bull. de la Société Impériale des Naturalistes de Moscou,' 1867, pt. 1, p. 121.

[5] "Die Lethæa Rossica und ihre Gegner., erster Nachtrag." Von E. von Eichwald. 'Bull. de la Soc. Imp. Nat. de Moscou,' 1867, iii, p. 202.

As the original specimen of *A. Eichwaldi* cannot be found, and it seems certain that two forms of pygidium have been observed and figured from Russia—one a *mucronate* and the other a *non-mucronate* form, it appears to be most satisfactory in every way to adopt the course suggested by von Möller, and give the name of *Phillipsia Eichwaldi*, Fisher sp , to the non-mucronate form, and retain for the other the name of *Ph. mucronata* of M'Coy.

As, however, it is equally certain from the evidence afforded by a large series of specimens—chiefly from the Scottish Carboniferous series—that there are *two forms*, apparently identical in every respect, *save* that one has a mucronate and the other a non-mucronate pygidium, it would be better to go a little further and propose to make M'Coy's species *mucronata* a variety of *Phillipsia Eichwaldi*, of Fischer.

Prof. Traquair retains the generic name of *Griffithides* for his mucronate form on the ground that the lateral glabellal furrows, characteristic of the typical *Phillipsiæ*, are wanting. We would venture to assert that, although not clearly to be seen in Prof. Traquair's specimens, they are present in all the examples of this species, and that when not visible it is simply due to the state of preservation of the individual specimen. The form is undoubtedly a true *Phillipsia ;* and, although at first we were unable to see the wisdom of attempting to maintain the two genera, *Phillipsia* and *Griffithides*, for these closely related Carboniferous Limestone forms of Trilobites, we are now prepared to maintain them without hesitation, and can point to good characters by which they may be readily distinguished. Of course, when the materials are very fragmentary it is next to impossible to define their place positively, but when once the whole specimen of any one species has been correctly figured (as we hope will be found to be the case in the plates accompanying this Monograph) a great deal of the difficulty disappears.

Not having been fortunate enough to see Mr. George Tate's type-specimen of *G. (Phillipsia) Farnensis*, we must accept as conclusive the opinion of Prof. Traquair that this species is only a synonym of *Phillipsia mucronata*, M'Coy. We have been favoured with the loan of the type-specimens of M'Coy's *G. mesotuberculatus*, from the Woodwardian Museum, Cambridge, and are satisfied that it is synonymous with his *Ph. mucronata*.

7. Phillipsia quadrilimba, *Phil.*, sp. 1836. Pl. VII, fig. 1.

Asaphus quadrilimbus, *Phillips*. Geol. Yorks., vol. ii, p. 239, pl. xxii, figs. 1, 2, 1836.

Phillipsia quadrilimba, *H. Woodw.* Cat. Brit. Foss. Crust., p. 56, 1877.

Nothing is known of this species save from Prof. Phillips' figure and description, which is here reproduced. I have not succeeded in tracing the original specimen.

The following is Prof. Phillips' description of *Asaphus quadrilimbus*, taken from the 'Geology of Yorkshire,' vol. ii, p. 239, pl. xxii, figs 1, 2.

"Fig. 1.—The head. Margin quadrato-carinate, minutely striated; surface smooth; eyes very minutely reticulated. Fig. 2.—Abdomen."

Prof. Phillips figures a portion of a detached head and an imperfect tail, but they certainly did not belong to the same individual, and certainly not to the same species.

Formation.—Top of Lower Scar Limestones.

Locality.—Bolland, Yorkshire.

Genus 2.—Griffithides, *Portlock,* 1843.

Outline oblong-oval; glabella pyriform, gibbous in front, destitute of lateral furrows; basal lobes inflated; cervical lobe broad; eyes small, lunate, smooth; axial furrow marking division of free cheek clearly defined, outline broadly triangular, outer posterior angle sometimes produced into a cheek-spine. Thorax with nine segments; pygidium rounded, composed of about thirteen coalesced somites.

The subjoined is the original description of the genus *Griffithides*, given by Portlock (1843), op. cit., p. 310.

"*Cephalothorax.*—Semi-oval, longitudinal; glabella strongly marked and gibbous, rounded in front, narrowed posteriorly into an obsolete neck, with a furrow more or less distinct on each side; *cheeks*, triangular spaces very slightly convex; *wings* either ending in an angle posteriorly or prolonged backwards in a flattened spine. Eyes near the axis, not large, lunate, smooth (?). The minute neck tubercle sometimes present.

"*Thorax.*—The pleuripedes are compound, in number nine, or with the neck-segment ten.

"*Pygidium.*—Fully developed and strongly resembling that of *Phillipsia.*

"A genus replacing *Asaphs* and *Phacops* in the Carboniferous system; it is dedicated to Mr. [afterwards Sir Richard] Griffith." Portlock also observes, p. 309:

"The determination of the true genus of the other Mountain-limestone Trilobites to be now described is attended with considerable difficulty, as no fully expanded specimen has as yet been found[1]; .... however, enough has been preserved to give ... a general idea of the forms of the cephalothorax and pygidium, and of the structure of the thoracic segments. The form of the cephalothorax, and the position and arrangement of the eyes, resemble closely the genus *Asaphus* as restricted, and parti-

[1] This was in 1843, but several complete specimens of *Griffithides* have since been obtained.

cularly *A. expansus* and *Hemicrypturus Razamowskii* of Green; but . . . . there is a perfectly developed pygidium, closely resembling that of *Phillipsia.* From the abundance of such pygidial relics, and the total absence of any other, it may fairly be presumed that a pygidium with axal and lateral segments fully developed was a generic peculiarity. The glabella is also more elevated than in the true *Asaphs.* In the absence of perfect specimens the formation of a new genus is difficult; but since the group is distinguished from the *Asaphs* of older rocks by the fully developed pygidium, it is advisable to separate them. It may also be here remarked that Green's cast of *Hemicrypturus Razamowskii,* and Pander's figures of *A. expansus,* seems to show that the lateral segments of the thorax in those species were compound, as in *Phacops* and *Calymene,* whereas they are merely folded in *Asaphus latifrons* of this work; in the species here figured they are also compound, so that there appears to be a blending of the characters of the several sections one into the other."

8. GRIFFITHIDES SEMINIFERUS, *Phillips,* sp. 1836. Pl. V, figs. 1—9; and Pl. VIII, fig. 14.

ASAPHUS SEMINIFERUS, *Phillips.* Geol. Yorks., vol. ii, p. 240, pl. xxii, figs. 8, 9, 10, 1836.
PHILLIPSIA GEMMULIFERA, *De Koninck.* Animaux Foss. Terr. Carb. Belgique, p. 603, pl. liii, fig. 3 (non Phillips), 1842.
— — *M'Coy.* Synopsis Carb. Foss. Ireland, p. 162, 1844.
— SEMINIFERA, *Morris.* Cat. Brit. Foss., p. 114, 1854
— — *Salter & H. Woodw.* Cat. and Chart Foss. Crust., p. 15, fig. 110, 1865.
— — *H. Woodw.* Cat. Brit. Foss. Crust., p. 55, 1877.

General form ovate-oblong: head-shield arcuate, glabella large gibbous, overhanging the anterior border; basal lobe pyriform; neck-lobe broad, separated by a wide furrow above and below; eyes small, reniform, smooth; raised portion of free cheek, glabella, and neck-furrow coarsely and irregularly granulated, margin of free cheeks smooth, lateral angles not produced into cheek-spines, thoracic segments nine, axis wider than pleuræ, only diminishing very slightly towards the pygidium; each segment ornamented by a single row of coarse granules (about eight on each side and ten on the axis); axal furrows strongly marked; segments arched, ends of pleuræ rounded, faceted portion smooth: pygidium composed of twelve coalesced somites, axis tapering gradually to an obtuse extremity; side ribs about nine, ornamented each by a single row of tubercles eight to ten on axis, about eight on each pleura; margin of pygidium narrow, edge bevelled.

Hypostome broad and short, wings not distinct from central lobe; obliquely striated, free extremity rounded and emarginated.

The subjoined is Phillips' description of *Asaphus* (*Griffithides*) *seminiferus*:[1]

"*Head* poroso-granulated, mesial lobe bisulcate on the sides, and bituberculated at the base; *abdomen* with tumid lobes; ribs roughened, with eight or ten unequal prominent subglobose puncta; limb not striated. The first segment of the middle lobe mucronate.

Formation.—"Rotten Stone" Band; Carboniferous Limestone.

Locality.—Matlock, Derbyshire; Settle, Yorkshire; and Blackrock, near Cork.

9. GRIFFITHIDES GLOBICEPS, *Phillips*, sp., 1836. Pl. VI, figs. 1, 3, 4, 5, 6.

ASAPHUS GLOBICEPS, *Phillips*. Geol. Yorks., vol. ii, p. 240, pl. xxii, figs. 16—20, 1836.
GRIFFITHIDES GLOBICEPS, *Portlock*. Rept. Geol. Londonderry, p. 311, t. xi, figs. 9 *a*, *b*, 1843.
PHILLIPSIA GLOBICEPS, *De Koninck*. Anim. Foss. Terr. Carb. Belgique, p. 599, tab. liii, fig. 1, 1844.
GRIFFITHIDES GLOBICEPS, *M'Coy*. Synopsis Carb. Foss. Ireland, p. 160, 1844.
— — *Oldham*. Journ. Geol. Soc. Dublin, vol. iii, pt. 3, p. 188, pl. 2. 1846.
— — *Morris*. Cat. Brit. Foss., p. 109, 1854.
— — *Salter & Woodw*. Cat. and Chart Foss. Crust., p. 16, fig. 117, 1865.
— — *H. Woodw*. Cat. Brit. Foss. Crust., p. 37, 1877.

General form ovate-oblong: head elevated, glabella very gibbous, overhanging the anterior border of shield, contracting rapidly to about half the width behind, where it unites with the neck-lobe; basal lobes prominent, triangular, to which the eyes seem to be united without the intervention of the fixed cheek; but there is a very narrow border united to the glabella forming the palpebral lobes which join the lateral lobes, or nearly so; neck-lobe narrow, axis strongly arched; lateral portion crossed by cheek-suture obliquely; eyes very small but exceedingly prominent; cheeks very narrow and compressed, ending in short, blunt spines; margin of head striated longitudinally; eyes very minutely faceted; thoracic segments nine in number, strongly trilobed; axis wider than the pleuræ; the posterior portion of each segment strongly corrugated, and each pleural groove extending rather beyond the fulcral point; the extremity of each pleura is rounded and broadly faceted: pygidium rounded, consisting of eleven coalesced somites, which in the axis continue the corrugated character of the thorax, but diminish to a

[1] See 'Geol. York.,' 1836, vol. ii, p. 240, pl. xxii, figs. 8, 9, 10.

blunt termination considerably within the border; the ribs of the pygidium are double and die out before reaching the edge of the tail-shield, leaving a somewhat wide smooth margin. Surface of head very finely punctate.

Formation.—Carboniferous Limestone.

Localities.—Bolland and Settle, Yorkshire; Forest of Wyre, Oreton; Milicent, Clane, Kildare; Waterford, Clonea; Derryloran, Tyrone; Blackrock, Cork; and Athlone, Ireland.

Hypostome.—A detached hypostome from Derryloran, Tyrone (see Pl. VI, fig. 5), belonging to one of these species of *Griffithides*, if not to *G. globiceps*, has very well-marked characters.

The upper border is strongly arched, the centre is tumid; the two wings form blunt angles, giving breadth to the attached anterior border; the sides curve inward almost to the lower end where there is a slight expansion; the lower free extremity is only half as wide as the upper; the rim or border is raised and the angles truncated, the inner portion of the lower extremity is slightly raised.

Of *Griffithides globiceps*, Prof. Phillips in his 'Geology of Yorkshire' (1836), vol. ii, p. 240, writes as follows:—"Limb quadrate, with four imbricating striæ; eyes lunate on a globular projection; head globular. (This agrees better than any other which I have seen with *E. Derbiensis* of Martin, t. 45, * 1.)"

General Portlock's description is as follows[1] (1843):—"*Glabella* short and almost globular in front; length four tenths, breadth three tenths of an inch; greatly elevated above the cheeks; narrowed at the base to less than one half the breadth in front; *cheeks* triangular, slightly convex; *eyes* short, lunate, connected with the glabella by a projection or nucleus; no visible reticulation; the *wings* end posteriorly in sharp angles, and in perfect specimens appear strongly striated; the margin is raised above the level of the cheeks; the neck-furrow is deep; total breadth of the cephalothorax more than seven tenths of an inch. This is referred to Professor Phillips' species, as it agrees closely with it in the characters of the cephalo-thorax, and is noted by him as occurring in the County of Kildare."

Dr. Oldham, in the 'Journ. Geol. Soc. Dublin,' 1846 (vol. iii, part 3, p. 188, pl. ii), figures and describes two very perfect specimens of *Griffithides globiceps*, which by the kindness of Prof. Hull, F.R.S., we are enabled to reproduce on our Plate VI, figs. 1 *a*, *b*, and 3.

Dr. Oldham speaks of this species as one of the most abundant and typical Trilobites of the Carboniferous Limestone." . . . General form, elongated, oval, body contractile, divided into three nearly equal parts by the cephalothorax, thorax, and pygidium; entire surface marked with minute irregularly disposed granulations, these are only seen in well-preserved specimens. *Cephalothorax* semi-elliptic; *glabella* short, pyriform, very tumescent, approaching to globular

[1] 'Geology of Londonderry,' p. 311.

in front, considerably elevated above the cheeks, narrowed behind to about one half its breadth in front, the tubercles, or projecting portions which connect the eyes with the glabella, forming the remainder of the breadth, divided from the cheeks by well-marked furrows; it is marked by three nearly obsolete cephalo-thoracic furrows on either side. These are scarcely seen when perfect, but are obvious in the cast; cast minutely punctured or granulated. Cheeks spherico-triangular, convex, so thickened on the outer edge as to form a distinct border or rim, elevated above the cheeks and rounded, which is prolonged backwards into short pointed spines (they are broken off in fig. 1). This rim or border ('wings') is marked with sharply raised longitudinal lines, the number varying in different specimens and in different parts of the rim, as they do not extend the whole length (our figures are deficient in showing this). These raised lines are wanting in the cast which is smooth. The raised rim is continued across the cephalo-thorax where it joins the thorax, and is here of the same form and structure as the first transversal segment of the thorax, from which it is sometimes not easily distinguished. In front of the glabella the rim is much smaller than along the margin of the wings, and is turned under the margin, forming a flat expansion. *Eyes* small, suboval, very prominent; when perfect covered with a smooth, corneous, transparent membrane; but under this very finely and beautifully reticulated, not oblique (the apparent obliquity of the eye being caused by the position of the tubercle or projection which unites it with the glabella), covered above with a distinctly granulated 'velum palpebrale,' the outline of which corresponds to the facial suture. Neck-tubercle frequently seen.

"In the furrows which separate the cheeks and glabella, about half way between the front of the eye and the anterior margin, I have observed in all the tolerably preserved specimens which I have seen, a small hole or indentation. These are constant and therefore obviously connected with the structure of the creature, although I cannot offer an explanation of their use. They are similar to those noticed by Portlock in his *Ampyx Sarsii*.[1]

"*Thorax*-joints nine in number; when perfectly preserved the joints of the medial segment or axis appear simple, but are marked internally with a transverse furrow, the joints of the lateral segments ('pleuripedes,' Portlock) are compound, being marked along their centre by a furrow which follows the outline of their form but does not reach the outer margin, and so formed by flattening on the edge as to admit of their folding freely over one another, when the animal was contracted; the axal and lateral lobes are nearly equal in breadth, but differ much in sectional form.

"*Pygidium*, a little more than semicircular, middle lobe consisting of eleven costæ, divided by well-marked furrows and simple (De Koninck says fourteen, our

[1] I refer to this structure in my Appendix to Part II.

specimens do not confirm this), which diminish successively; the lateral lobes have about thirteen costæ, simple and united at the margin by a smooth rim, the furrows which divide the costæ becoming obsolete or nearly so before they reach the outer edge; this smooth rim occupies about one third of the breadth of the lobe.

"When rolled up this smooth rim partly covers the wings of the thorax. .... It is remarkable that M. de Koninck, in his excellent 'Description des Anim. foss. dans le terr. Carbonif. de Belgique,' though figuring (pl. liii, fig. 1) a Trilobite with strongly reticulated eyes, as *P. globiceps*, states in the text, in more than one place and most particularly, that the eyes are smooth. Indeed, although his description is in general accurate enough, his plate represents a fossil, which in many respects totally differs from the *Asaphus globiceps* of Phillips to which he refers it."

10. Griffithides acanthiceps, *H. Woodw*, *sp. nov.* Pl. VI, figs. 2, 10, and 11; and Pl. VII, figs. 2 and 3.

Head-shield semicircular, produced in front, glabella very gibbous, overhanging the anterior margin, twice as wide in front as at the nuchal furrow, the whole surface strongly granulated; basal lobes very small, rounded; neck-furrow deep; neck-lobe rounded; fixed cheeks exceedingly narrow, scarcely discernible, forming a rounded palpebral lobe (the surface of which is granulated) over each eye and a narrow rim around the glabella; eyes small, finely faceted; inner raised portion of cheeks granulated, margin smooth, posterior angle produced into long cheek-spines equal to the glabella in length.

Thorax consisting of nine free segments, surface of thorax smooth without ornamentation; axis arched, rather wider than its pleuræ, broader next the head and diminishing very slowly to the pygidium; each of the pleuræ strongly grooved down the centre, posterior portion rounded and slightly raised, anterior portion slightly depressed; fulcral points distinctly marked, extremity of pleuræ faceted in front and rounded.

Abdomen or pygidium composed of about thirteen coalesced somites, border smooth, slightly channelled, ribs terminating close to border. Extremity of pygidium very slightly pointed.

The specimens in which the head, thorax, and abdomen of this species are preserved united, are figured on Pl. VII, figs. 2 and 3, having been discovered too late for insertion on Pl. VI.

Formation.—Carboniferous Limestone.

Localities.—Craco, near Grassington; Settle, Yorkshire; and Castleton, Derbyshire.

I had certainly no intention of burdening the list of Carboniferous Limestone Trilobites with another species, and I had, in fact, placed figs. 2, 10, and 11, Pl. VI, provisionally under *G. globiceps;* but on a more detailed examination of the original of fig. 10, it occurred to me something more might be made out by developing it further. To my surprise two long spines were uncovered, and I then perceived that figs. 2 and 11 had at one time also possessed cheek-spines, but these have since been broken off. *G. globiceps*, on the contrary, has blunt and short cheek-spines, large basal lobes to the glabella, and the portion of the test preserved shows the head-shield to have been finely punctate, whereas in *G. acanthiceps* the surface of the head-shield is distinctly granulated.

The discovery of the detached head with the long cheek-spine of this species, proving it to be distinct from *G. globiceps*, led me to a further and closer examination among the remaining doubtful specimens, and one belonging to Mr. J. Aitken, having been skilfully developed by our (Brit. Mus.) Mason, Mr. C. Barlow, revealed the same character of the head as in the specimen from the Woodwardian Museum, and in addition exposed the thorax and abdomen very fairly preserved, enabling me to complete the description of *G. acanthiceps* (see Pl. VII, fig. 2).

Another specimen, also from Cambridge, exhibiting a detached head and pygidium upon the same piece of matrix, is drawn (on Pl. VII fig. 3).

11. GRIFFITHIDES LONGICEPS, *Portlock*, 1843. Pl. VI, figs. 7, 8, and 9.

GRIFFITHIDES LONGICEPS, *Portlock*. Rep. Geol. Londonderry, p. 310, t. xi, figs. 7 *a*, *b*, 1843.
— — *M'Coy*. Synopsis Carb. Foss. Ireland, p. 160, 1844.
— — *Morris (in part)*. Cat. Brit. Foss., p. 109, 1854.
— — *H. Woodw. (in part)*. Cat. Brit. Foss. Crust., p. 37, 1877.

General form ovate-oblong; head-shield very large in proportion to the rest of the body, forming two fifths of the entire length; glabella very gibbous, pyriform, basal lobes obtusely triangular, with a tubercle on the centre of each; fixed cheeks very narrow, but expanding rather at the sides of the glabella in front of the eyes; axal portion of the neck-lobe very broad, and separated by a strong furrow, and bearing one tubercle on its centre; eyes moderately large, reniform, surface very finely faceted; raised inner portion of free cheek rather narrow, surface finely granulated, outer margin wide, posterior angles produced into broad and stout spines, reaching to the fifth segment of the thorax; thorax composed of nine free segments, the axis arched, equalling half the entire breadth of the thorax; each segment bordered by ten or eleven granules on its axis along the posterior border, and seven or eight on each pleura; pleuræ rounded at their extremities, pygidium composed of thirteen coalesced somites, ornamented in a similar manner to the free thoracic ones ; axis tapering to a blunt extremity, and surrounded at its termination by the smooth border of the tail-shield; ribs nine in number, dying out near the margin.

Formation.—Carboniferous Limestone.

Localities.—Settle, Yorkshire; Cookstown, Tyrone; Creggane, Limerick; Brockley, near Lesmahagow.

The following is Portlock's original description of his *Griffithides longiceps* :[1]

" *Glabella* elevated, rounded, and swollen in front, but narrowing gradually towards the base, the external surface covered with minute dots, and, when removed, the surface below appears rather granular or rugose; length rather more than four tenths of an inch, breadth in front three tenths, and at the base three twentieths; eye-projection, very small, and near the base. Thorax imperfect, but showing the lateral segments to be compound. Pygidium, in length four tenths, breadth five tenths of an inch; axal lobe elevated; segments thirteen, with about twelve granules or small tubercles on each, and a rounded extremity also covered with granules; lateral segments about nine on each side, with a row of small tubercles on each, not quite extending to the margin, which is flattened or turned up, and strongly striated on the edge and under surface. The prolongation of the segments is marked by slight elevations on the margin; the flattened and turned-up margin of the pygidia of this genus is generally more striking than in *Phillipsia*, in which the pygidial margin follows the slope of the lateral segments, but the character depends in part upon the condition of the crust, being more marked when it is decorticated."

In Prof. Morris's 'Catalogue of British Fossils,' p. 109, and also in H. Woodward's 'Catalogue of British Fossil Crustacea,' p. 37, Portlock's *Griffithides longispinus* ('Geol. Rep.,' Lond., p. 312, pl. xxiv, fig. 12) is made a synonym of *Griffithides longiceps*, Portl. A careful examination of Portlock's type specimen, kindly lent to me by the authorities of the Museum of Practical Geology, Jermyn Street, enables me to state that it is quite distinct from *G. longiceps*, and I figure it, with others, on Pl. VII, figs. 5 and 6.

12. GRIFFITHIDES PLATYCEPS, *Portlock*, 1843. Pl. VI, fig. 13.

GRIFFITHIDES PLATYCEPS, *Portlock*. Rep. Geol. Lond., p. 311, pl. xi, fig. 8, 1843.
— — *Morris*. Cat. Brit. Foss., p. 109, 1854.
— — *H. Woodw*. Cat. Brit. Foss. Crust., p. 37, 1877.

[1] 'Geology of Londonderry' (1843), p. 310.

This species, which was founded upon a glabella only by Captain (afterwards General) Portlock, who obtained it from the Carboniferous Limestone of Tyrone, Ireland, is thus noticed by him in his Report on the Geology of Londonderry and Tyrone' (p. 311). "Fig. 8 is a larger individual, of probably another species [distinct from *longiceps*]; the surface is granular, and it is proportionately flatter; it may be called *Griffithides platyceps*." The specimen figured on our Pl. VI, fig. 13, and enlarged twice the natural size, was obtained from the Carboniferous Limestone of Derryloran, Tyrone, the original being preserved in the Museum of the Geological Survey of Ireland, Dublin.

13. GRIFFITHIDES OBSOLETUS, *Phillips*, sp., 1836. Pl. VI, fig. 12.

ASAPHUS OBSOLETUS, *Phillips*. Geol. Yorks., vol. ii, p. 239, pl. xxii, figs. 3—6, 1836.
— GRANULIFERUS, *Phillips*. *Ibid.*, fig. 7, 1836.
PHILLIPSIA BRONGNIARTI, *De Koninck*. Anim. Foss., t. liii, fig. 7, 1842.
— — *Morris*. Cat. Brit. Foss., p. 114, 1854.
— — *Salter & H. Woodw.* Cat. and Chart Foss. Crust., p. 16, fig. 113, 1865.
— — *H. Woodw.* Cat. Brit. Foss. Crust., p. 55, 1877.

This species is founded on a very broad and smooth pygidium, nearly one fourth broader than long, composed of ten coalesced somites, the axis much broader than the pleural portion; each of the nine rib-like plicæ marked by a furrow down the centre (as in the pygidium of *G. globiceps* already noticed), the margin of the tail-shield is smooth.

The glabellal portion of the head, most probably belonging to the same individual (being enclosed in the same piece of matrix), although mutilated, exhibits peculiar and delicate striations over its entire surface. The head (somewhat restored) is figured by Prof. Phillips with the pygidium. This specimen is the type of Phillips' figure in the 'Geol. of Yorkshire,' and was at that time in the Gilbertson Collection, and is now in the British Museum (Natural History).

Formation.—Carboniferous Limestone.

Locality.—Bolland, Yorkshire.

The following is Prof. Phillips' original description of his *Asaphus obsoletus* (op. cit., p. 239):

"Abdominal lobes ventricose; transverse undulations obtuse; surface smooth with undulating lines; *the limb with oblique undulating striæ;* head finely striated, undulated lines roundish and lumpy."

The name of *Ph. Brongniarti*, Fischer, sp., 1825, being the oldest, had been adopted by De Koninck in 1841 for this specimen in lieu of Phillips' name of *obsoletus* (given in 1836), which has since figured as a synonym; but as Fischer de Waldheim expressly says in his work,[1] already quoted in reference to two pygidia of Trilobites named by him, *Asaphus Eichwaldi* and *A. Brongniarti*, "I think that size alone is not sufficient to make two species; I consider them as one and the same species, for which I have retained the name of Eichwald, the more so as another Trilobite already bears the name of Brongniart." It seems, therefore, clearly undesirable to revive a specific name which its author had already cancelled, and to apply it to a form which certainly cannot be correlated with that originally intended to be described under the defunct term.

14. GRIFFITHIDES LONGISPINUS, *Portlock*, 1843. Pl. VII, figs. 5 *a*, *b*, *c*, and 6.

GRIFFITHIDES LONGISPINUS, *Portlock*. Geol. Rept. Lond., p. 312, pl. xxiv, fig. 12, 1843.
— — *M'Coy*. Carb. Foss. Irel., p. 161, 1844.
— LONGICEPS, *Morris*. Cat. Brit. Foss., p. 109, 1854.
— — *Salter and H. Woodwood*. Cat. and Chart Brit. Foss. Crust., p. 16, fig. 115, 1865.
PHILLIPSIA — *V. von Möller*. Trilob. der Steinkohl., pp. 19 and 73, 1867.
GRIFFITHIDES — *H. Woodward*. Cat. Brit. Foss. Crust., p. 37, 1877.

General form elongated-oval; head wider than long; glabella very gibbous in front, slightly overhanging the anterior border, much broader in front than behind the eyes; basal lobes small, rounded; neck-lobe strongly arched, narrow, divided from the glabella by a deep neck-furrow; fixed cheeks narrow, where they pass from the posterior border and above the eyes, forming the small, rounded, palpebral lobes, after which they expand again slightly on each side of the glabella before the facial suture unites with the front border; surface of glabella thinly and irregularly tuberculated; free cheeks small, elevated, channelled around the eye and the border, the small area so enclosed covered with numerous, rather coarse, and irregular bead-like ornamentations; eyes reniform, moderately small, smooth; margin of cheeks produced into rather long cheek-spines ("long, flat, striated spines." Portlock[2]); margin of head-shield incurved and finely striated. Thoracic segments nine in number, axis strongly arched, each segment having a narrow elevated central rib, ornamented with about twelve small tubercles or spines, with a smooth anterior articular portion and a less elevated posterior border; the pleuræ are strongly grooved, and are bent down at the fulcral point, their

[1] 'Oryc. Gouv. Moscou,' p. 121.

[2] There is reason to conclude that one of these spines existed when Portlock wrote his description, although it is only now indicated by a fragment and by the scar where it once rested.

extremities being faceted and obtusely pointed. Pygidium composed of from twelve to fifteen[1] coalesced segments; axis strongly arched and ribbed like the thorax, but no ornamentation visible; side lobes of pygidium also arched; the ribs running to the border; a wide striated margin is exposed where decorticated.

Formation.—Carboniferous Limestone.

Locality.—Carnteel, Tyrone, Ireland.

I was formerly of the same opinion as my friend Prof. Morris that *Griffithides longispinus* of Portlock was identical with the *G. longiceps* of the same author, and this opinion was also shared by V. von Möller; but, having been favoured with the opportunity of examining Portlock's original specimens of both these forms, I have satisfied myself that they are entitled to be kept distinct.

In *G. longispinus* the small raised cheeks are covered with rather large bead-like tubercles; the glabella is longer in proportion, and has a few scattered tubercles on the surface; the neck-lobe is narrow.

In *G. longiceps* the head-shield is finely granulated, the neck-lobe is broad, the basal lobes are larger, and only one tubercle marks the centre of each. The thoracic segments are more strongly arched in *G. longispinus*, and the pygidium is composed of more somites than in *G. longiceps*, and is more elongated and more highly arched than in the latter.

I think a comparison of our figures of *G. longiceps* (Plate VI, figs. 7 and 8) with those of *G. longispinus* (Plate VII, figs. 5 *a*, *b*, *c*, and 6) will satisfactorily show all the points in which these two species resemble and differ from one another.

Subjoined we give Portlock's original description of his *Griffithides longispinus*, 'Report on the Geology of Londonderry,' &c., p. 312, 1843:

"This beautiful species approximates to *G. longiceps*, as the specimen figured in pl. xi, fig. 9, is imperfect, and may, therefore, have had lateral spines; there are, however, some good marks of distinction, the glabella is not quite so long nor so narrow, and the pygidium is rather longer, has fifteen axal segments, and is smooth, whilst the thoracic segments are granulated; these may, however, be only in part accidental variations from peculiar circumstances, wear, &c. The glabella is gibbous, and longer than in *G. globiceps*. The wings [free cheeks] pass anteriorly under or in front of the glabella, and are prolonged backwards in long, flat, striated spines (see magnified view, pl. xxiv, fig. 12 *b*). Though the granulations are well marked on the thoracic segments (pl. xxiv, fig. 12 *e*) none are visible in this specimen on the pygidium (fig. 12 *c*); when decorticated a large striated and nearly vertical margin appears, as in fig. 12 *d*, which also shows that the under surface does not exhibit the extension of the segments to the margin

[1] The extremity is injured so that the exact number cannot now be ascertained.

which is seen on the upper surface of the crust, and hence that in the two conditions they might appear different fossils."

15. Griffithides calcaratus, *M'Coy*, sp., 1844. Pl. VII, fig. 13.

Griffithides calcaratus, *M'Coy*. Synop. Carb. Foss. Ireland, p. 160, pl. iv, fig. 3, 1844.
— — *V. von Möller*. Trilob. der Steinkohl., p. 19, 1867.
— mucronatus, *H. Woodw.* Cat. Brit. Foss. Crust., p. 37, 1877.

This species was founded by Prof. M'Coy in 1844 for a specimen from Ireland,[1] of which he figures the head only, but describes the head and tail also. We have not been so fortunate as to see the original of M'Coy's figure, but we give his own description as follows:

"Cephalothorax semi-oval; glabella smooth, ovate, most convex in the middle of its length; cheeks small, triangular, flat, smooth; wings strongly striated, broad, prominent, rounded, terminating posteriorly in long flattened spines; eyes moderately lunate (smooth?), connected with the glabella by a nucleus on each side; pygidium with a smooth margin, each segment with a row of very minute granulations.

"This beautiful species is most nearly allied to the *G. longispinus* of Portlock, but is at once distinguished by its smooth cheeks; the eyes, also, in the present species, are differently formed and placed, and the glabella is much smaller and less prominent in front. Length of glabella five lines; greatest width three lines; width at base one line; width of cephalothorax seven lines; length of eyes one and a half lines; width one line; length of posterior alar spine three lines.

"The pygidium has a broad, smooth margin or limb, in which it differs from that of *G. longispinus*, in which the segments are extended to the margin. There is a single row of very minute granules on each segment. Width of pygidium five lines" (op. cit., p. 160).

Although I placed *G. calcaratus* in 1877 as a synonym under *G. mucronatus*, I find that neither the description nor figure admit of its being so disposed of, and I therefore give it on M'Coy's authority.

Valerian von Möller says of *Griffithides calcaratus:* "M'Coy only figures the cephalothorax. It is nearly related to *Ph. globiceps*, Phill., but is distinguished from this species by the very inferior size of the glabella, and by the long, flattened cheek-spines; and from another Trilobite, *Ph. longiceps*, Portl. (= *G. longispinus*, Portl.), with which it is also nearly connected, it differs in its very narrow glabella and the wide flattened border around the pygidium (*op. cit.*, p. 19)."

[1] Said to be from the Upper Limestone, Roughan, Dungannon.

16. Griffithides brevispinus, *H. Woodw., sp. nov.* Pl. VII, figs. 7, 8.

Among the various specimens received from Mr. Robert Craig, of Langside, Beith, Ayrshire, N.B., are two fragments of heads of a small species of Trilobite of the genus *Griffithides*, which, as the discoverer observes, appear to differ from any which have hitherto come under notice from the Carboniferous Limestone. The head is nearly twice as broad as it is long, the free cheek terminating laterally in a short spine; the eye, which is very smooth, is rounder and more tumid than in other species, and the facets, which are discernible with a Browning's platyscopic lens, are very minute, and do not break the smooth hyaline surface of the compound eye.

The glabella is nearly smooth in front, and overhangs the anterior border of the head-shield; the posterior portion of the glabella and the neck-lobe are irregularly tuberculated. The free-cheek is also tuberculated, and has about eight tubercles on each cheek, placed in a semicircle around the compound eyes. The margin of the shield is raised and striated, and has a rather deep and smooth furrow between the raised border and the inner portion of the free-cheek.

In the style of its ornamentation this form agrees most nearly with *G. longispinus*, Portl., but the head is shorter and broader, and the spines are only one half the length of that species. I have named this species *G. brevispinus*, H. Woodw.

Formation.—From the Lower Carboniferous Limestone.

Locality.—Langside, Beith, Ayrshire.

From the collection of Mr. Robert Craig, of Langside, Beith, Ayrshire, N.B.

17. Griffithides moriceps, *H. Woodw.*, 1883. Pl. VII, figs. 9, 10, 11, 12.

Griffithides moriceps, *H. Woodw.* Geol. Mag., Dec. ii, vol. x, p. 487, pl. xiii, fig. 4, 1883.

Head-shield semicircular, 22 mm. long and 30 mm. broad; glabella elevated, very gibbous and obtuse in front, and twice as wide at the anterior border as at the neck-furrow; basal-lobe small, rounded; surface of glabella and free-cheeks thickly covered with large round granulations; facial suture running very close around the glabella; free-cheeks hatchet-shaped, central part ornamented and raised, bearing the small smooth reniform eyes close to the basal lobe; border of cheeks deeply furrowed and smooth, with a raised margin having a rounded rim; neck-lobe narrow, smooth, flattened, and separated by a rather strongly marked neck-furrow, which is continued and unites with the furrow surrounding the free-cheeks, forming somewhat blunt angles to the head-shield.

Formation.—Carboniferous Limestone.

Locality.—Settle, Yorkshire.

Only the head of this rather large species is known, but it differs in many respects so markedly from its nearest ally, *G. seminifera*, that I feel justified in making it a distinct species.

The glabella is broad and rather obtuse in front and very narrow behind; the basal lobes are small and drop-shaped; the eyes are very small, placed far back on the cheeks and near the basal lobes; the granulations on the head and cheeks are coarse and closely placed, and suggested the specific name. Along the anterior border of the head these granulations are elongated, and form a close and regular marginal ornamentation (see Pl. VII, fig. 11). There are five specimens (all detached heads) of this species, from the Carboniferous Limestone of Settle, Yorkshire. They are all preserved in the Woodwardian Museum, Cambridge.

18. Griffithides glaber, *H. Woodw., sp. nov.* Pl. IX, figs. 4 *a* and 4 *b*.

I am indebted to Mr. E. T. Newton, F.G.S., Assistant Naturalist to H.M. Geological Survey of Great Britain, for drawing my attention to this new and interesting form of Carboniferous Trilobite, which I had not observed in the collection on my former visits to the Museum of Practical Geology, Jermyn Street.

Both the specimens figured are preserved in a dark crystalline rock from the Carboniferous Limestone, Castle-Mumbles, Glamorganshire. The extreme length of the more complete specimen (fig. 4 *a*) is 36 mm., and its greatest breadth 13 mm. Length of glabella, including neck-lobe, 12 mm.; length of pygidium 13 mm. The head is much mutilated, but sufficient of it remains to show that the glabella was smooth, rather tumid, longer than wide, with a basal-lobe on each side near the neck; neck-lobe moderately broad, and marked with one central tubercle; the eyes are not preserved; border of free-cheek terminating in a short lateral spine on each side, and striated below. Free thoracic somites nine in number; extremities of pleuræ smooth and truncated; axis rather wider than lateral portion of somite; coalesced abdominal somites about nine in number; axis nearly smooth, with a slightly serrated posterior edge to each body-ring, margin smooth, rather broad; the pygidium is somewhat narrower and more elongated than in the other Carboniferous species.

Five specimens of pygidia referable to this species are preserved in the Museum of Practical Geology, from the Carboniferous Limestone of Northumberland, and a sixth specimen, in the same collection, from the Upper Carboniferous Shale of Ashford, Derbyshire.

This species in general form agrees most nearly with *G. longispinus*, Portl. (see *ante*, p. 36, Pl. VII, figs. 5 *a*, *b*, *c*, and 6), but *G. glaber*, as its name implies, is smooth or nearly so, whereas *G. longispinus* is coarsely tuberculated both on the head and body segments, the tail alone being smooth.

In *G. glaber* there is a trace of minute serration seen on a pygidium from Ashford, Derbyshire, not discernible on the other specimens, but I consider that they all belong to one and the same species.

Formation.—Carboniferous Limestone.

Localities.—Castle-Mumbles, Glamorganshire; Northumberland; and Ashford, Derbyshire.

All the above specimens are preserved in the Museum of Practical Geology, Jermyn Street.

19. Griffithides? Carringtonensis, *Eth.*, MS. Pl. IX, figs. 6 *a* and *b*.

This species, named by Mr. Etheridge, F.R.S., in manuscript, is represented by two pygidia in the collection of the Museum of Practical Geology, Jermyn Street, and one in British Museum series.

Head-shield and thorax unknown. *Abdomen* or *Pygidium.*—The largest specimen (which is preserved in the British Museum) measures 17 mm. in breadth and 15 mm. in length; the axis is 7 mm. broad at its proximal end, diminishing to $3\frac{1}{2}$ mm. at the distal extremity. It terminates at a distance of 3 mm. from the posterior border, which is smooth, and continues so around the semicircular margin of the pygidium. Twelve coalesced somites are indicated by as many broad and flattened rings in the axis, which have a faint vertical line crossing them on each side of the axis and parallel to the axal furrows. The pleuræ, nine in number, terminate abruptly about 3 mm. from the margin, and are each divided by a median groove.

The other and smaller specimens are fig. 6 *a*, measuring 10 mm. wide by 7 mm. long, and fig. 6 *b* 11 mm. wide by 8 mm. long.

This species agrees most nearly with *G. obsoletus*, but the latter has no distinct margin to the pygidium, and has only ten axal rings.

Formation.—In white crystalline Carboniferous Limestone.

Localities.—Falls Brew, Caldbeck, Cumberland (Mus. Brit.). Fig. 6 *a*, Longnor, Derbyshire, and fig. 6 *b*, from Narrowdale. (Mus. Pract. Geol.)

Note on GRIFFITHIDES LONGISPINUS, *Portlock*, Pl. IX, fig. 3. (See *ante*, p. 36.)

The specimen figured on Pl. IX, fig. 3, represents a detached glabella of this species, with a hypostome lying close to its posterior border. The glabella is also of interest because it exhibits (as a cast in relief), on the side of the decorticated head, the cast of the preocular pore (*p*), so well seen in *Griffithides globiceps* and *Ph. Eichwaldi* (Pl. IV, fig. 6, 8, 10, and 15).

Hypostome.—The anterior border of the hypostome measures 7 mm. in width, and it is also 7 mm. long. The posterior border is narrow and pointed. The middle of the hypostome is very strongly curved; the surface, where preserved, shows it to have been ornamented with delicate branching and wavy raised lines or striæ. Two muscular impressions or indentations mark the hypostome near the narrow extremity. The margin of the hypostome appears to have been thickened.

Formation.—Carboniferous Limestone.

Locality.—Longnor, Derbyshire.

Original specimen in the Museum of Practical Geology, Jermyn Street.

*** The following three species, referred to the genus *Phillipsia*, have been recognised since pp. 11—27 were in print, and require description.

20. PHILLIPSIA LATICAUDATA, *H. Woodw.*, *sp. nov.* Pl. VII, fig. 4.

Head imperfect; glabella tumid, rounded in front, with a narrow, smooth, raised marginal rim; general surface smooth, but finely punctated under a lens; basal lobe separated by a deep semicircular furrow from the rest of the glabella, and with two short lateral furrows on each side. Neck-furrow deep; neck-lobe rounded with one prominent tubercle on the centre. Length of glabella 5 mm., breadth $3\frac{1}{2}$ mm., cheeks not preserved.

Free thoracic segments unknown. Pygidium much broader than long, very strongly trilobed; axis elevated, consisting of twelve coalesced somites; each ring very strongly ridged, and each ridge ornamented with a line of minute tubercles; side pleuræ nine in number, rather broad for half their length, and minutely ornamented, but becoming fainter for the latter half, and dying away near the margin, which is almost smooth. Length of pygidium 6 mm., of axis 5 mm., breadth of tail 9 mm., breadth of axis 4 mm.

Formation.—Carboniferous Limestone.

Locality.—Bolland, Yorkshire.

There are four examples of this small species in black Carboniferous Limestone from Bolland, part of the Gilbertson collection preserved in the British Museum (Natural History). Each specimen of a pygidium has also a detached glabella preserved together in the same piece of matrix. There seems no doubt that the heads and tails originally belonged to the same individuals. They are quite unlike the ornamented pygidia of other species of *Phillipsia*, being broader and shorter, and more delicately ornamented, the pleuræ of the tail in particular being very peculiarly marked in their decoration and form, and in the break in their character midway. They appear to be worthy of specific recognition.

21. Phillipsia scabra, *H. Woodw.*, *sp. nov.* Pl. IX, figs. 5 *a*, *b*.

This species is based upon a head-shield and two pygidia from the Carboniferous Limestone, Vallis Vale, Frome, Somerset, preserved in the Museum of Practical Geology, Jermyn Street.

These are the first remains of Trilobites I have received from the Carboniferous Limestone of the south-west of England.[1]

The head-shield is 12 mm. in breadth by 8 mm. long. The glabella is prominent, rounded in front, but not overhanging the raised anterior border of the cephalon; three oblique furrows mark the sides of the glabella, the front furrow being nearly in a line with the anterior angle of the eye; the basal lobe is large, obtusely triangular in form; neck-lobe 1 mm. deep, divided by a shallow furrow from the glabella; pleuræ of neck-lobe extending for three-fourths the breadth of free-cheeks, and terminating acutely along their posterior margin; lateral border of glabella narrow, but expanding into a moderately broad margin in

[1] Mr. R. H. Valpy, F.G.S., Enborne Lodge, Newbury, informs me that he discovered a bed of shale of Carboniferous age with Trilobites in an excavation made for the Rifle-butts on the top of Black Down, on the Mendip Hills, Somersetshire. They were associated with Entomostraca.

front of the glabella. Free-cheeks small, with a broad and very distinct margin separated by a deep furrow; the margin is striated longitudinally; the head and cheeks are scabrous, most strongly so on the posterior half of the glabella. The eyes were large, nearly 3 mm. in length; they are unfortunately wanting, being represented by the cavity only.

Thoracic segments unknown; probably nine in number.

Pygidium.—The coalesced series of abdominal segments forming the pygidium are about fifteen in number, measuring 8½ mm. long by 10 mm. in breadth. The axis of the tail is broad at the proximal end, and roundly elevated; it decreases in size somewhat rapidly towards the posterior border which it overlaps. The margin of the pygidium is smooth for the breadth of one millimètre.

Formation.—In light reddish-brown coloured shale of Carboniferous age.

Locality.—Vallis Vale, Frome, Somerset.

Phillipsia scabra approaches most nearly to *Ph. gemmulifera* in general appearance, but in the latter species the glabella is smooth, and has only two oblique furrows on its sides; the neck furrow is smooth in *P. gemmulifera*, but finely tuberculated in *P. scabra.*

The pygidium in both species shows a smooth margin, which is widest in *P. scabra.*

The only specimens I have seen of this species are in the Museum of Practical Geology.

22. Phillipsia carinata, *Salter*, MS. Plate IX, fig. 7.

This species, named in MS. by the late Mr. Salter, is represented by two pygidia in the collection of the Museum of Practical Geology, Jermyn Street.

It owes its trivial name to the fact that the axis of the tail is acutely ridged, not roundly arched as in most of the other species. At first sight this might be supposed to be the result of crushing, but a closer examination shows that this is not the case, both specimens being similarly ridged.

The specimen figured (Pl. IX, fig. 7) measures 12 mm. broad by 9½ mm. in length, breadth of axis at the proximal border 5 mm., at the distal extremity 2 mm., length of axis 8½ mm. There are seventeen coalesced rings in the axis, and ten pleuræ on each side. Most of the surface has been decorticated, but where the shelly crust is preserved we see that each ring is ornamented by a single row of small tubercles placed rather far apart.

In Portlock's Geology of Londonderry and Tyrone, on pl. xi, fig. 10, there is a pygidium figured which seems to have been intended for a caudal shield of this very species; but the author makes no allusion to the figure in his text, nor yet in the explanation to the plate.

Formation.—Carboniferous Limestone.

Locality.—Derbyshire.

This species most nearly resembles the pygidium of *Ph. truncatula*, Phil. sp., but the acutely-ridged character of the axis in *P. carinata* suffices to distinguish it from this and other species.

Note on the Synonymy of PHILLIPSIA GEMMULIFERA, *Phillips, sp.*, 1836. (See *ante*, p. 17.)

Although the Carboniferous Trilobites are but few in number, and are all included in four genera, they have not escaped the usual trouble arising from incorrect determinations.

One of these occurred in reference to *Phillipsia gemmulifera*, Phillips, sp., better known by the name of "*Phillipsia pustulata*," Schlotheim, sp., a name applied to this form by Professor de Koninck in 1842-44 (see 'Descr. Anim. Foss. Terr. Carbonif. de Belgique,' p. 603, tab. liii, fig. 5).

This Trilobite, first known by a pygidiun only, was very carefully figured in Brongniart's and Desmarest's 'Histoire Naturelle des Crust. Foss.,' 1822, pl. iv, fig. 12, p. 145, where it is called "*Asaphus*" from the Black Limestone in the environs of Dublin.

It was next figured by Phillips in his 'Geology of Yorkshire,' 1836, vol. ii, pl. xxii, fig. 11, p. 240, who named it *Asaphus gemmuliferus*.

Buckland again repeated the figure later in the same year, and followed Phillips' name of *A. gemmuliferus*.

Professor de Koninck, in 1842-44, changed the name to *Phillipsia pustulata*, quoting Schlotheim's 'Nachtrage zur Petrefactenkünde' (ii Abth., pp. 42-3, Gotha, 1823, and 'Atlas,' p. 22, and plate xxii, fig. 6) as his authority. Now, as Schlotheim's Trilobite differs very greatly from Phillips's figure, and also from those given by Brongniart and by Buckland, and as, moreover, Schlotheim's specimen was said to have been derived from the youngest Upper Transitional Limestone (Devonian) of the Eifel, I felt great doubt in accepting Professor de Koninck's correlation of *Phillipsia gemmulifera* with *Trilobites pustulatus* of Schlotheim.

These doubts I expressed in a previous part of this Monograph (see pp. 17—19,

Pl. III, figs. 1—8), and also in the 'Geological Magazine' (Decade ii, vol. x, p. 450), and acting on the evidence of the age (Devonian), and the published figure of Schlotheim's specimen, I restored Phillips's specific name of *gemmulifera* for this Carboniferous Limestone form, and discarded that of Schlotheim (*T. pustulatus*) as untenable. Having lately seen and consulted my friend Prof. Dr. Ferdinand Roemer, of the Mineralogischen Museum, Breslau, our highest authority on the fossils of the Eifel, he very kindly promised me, on his return journey, in passing through Berlin to Breslau, to compare my figure of *Phillipsia gemmulifera* with Schlotheim's specimen of *Trilobites pustulatus* in the Berlin Museum. I now have the pleasure to append his letter, which entirely sets the matter at rest.

"*Mineralogical Museum of the*
"*Royal University of Breslau.*

"Dear Dr. Woodward,

"Schlotheim's *Trilobites pustulatus* is nothing else than a pygidium of *Phacops latifrons* from the Eifel. This is proved beyond any doubt by Schlotheim's original specimen in the Berlin Museum.

"Yours very truly,
"Ferd. Roemer."

"17th October, 1883."

Genus 3.—Brachymetopus, *M'Coy*, 1847.

General form elliptical; head-shield semicircular and slightly pointed, about one third wider than long; glabella small, somewhat elevated, one third the width of the entire shield and about one half the length, having a basal lobe on each side, but no short lateral furrows on the glabella; neck-furrow distinctly marked, equal in width to the posterior border of free cheeks; eyes small, smooth, equal to half the length of the glabella; no facial suture visible, only the axal-furrow surrounding the glabella and the neck-furrow; free-cheeks slightly convex, nearly twice as long as they are broad, with no visible suture separating them from one another in front of the glabella, margin broad and slightly grooved, angles of cheeks produced posteriorly into spines. The entire surface of the head covered irregularly with a small bead-like ornamentation.

Thoracic segments not known, probably nine in number.

Pygidium consisting of a variable number of segments, from ten to seventeen according to the species, the axis tapering rapidly to a bluntly-rounded extremity, each segment of axis ornamented with bead-like granulations, ribs with a double furrow extending nearly to the border which is smooth and rounded.

The first specimen of this genus was obtained by Mr. Frederick M'Coy, from the Carboniferous Limestone of Kildare and transmitted to Portlock, who, whilst struck by its dissimilarity to other Mountain Limestone forms, with which he was familiar, placed it provisionally in the genus *Phillipsia*, and described it as *Phillipsia Maccoyi* (see his 'Geology of Londonderry,' &c., 1843, p. 309, pl. xi, fig. 6).

It was left to Prof. M'Coy, in 1847, to propose a new genus for this and two other species, one of which is from Australia.

We append Prof. M'Coy's original description of the genus, from the 'Annals and Magazine of Natural History,' 1847, vol. xx, p. 230.

"*Brachymetopus* (M'Coy), new genus, 1847 (Etym. βραχύς, short, and μέτωπον, *the forehead or glabella*).

"*Gen. Char.*—Cephalothorax truncato-orbicular; limb (free-cheek) narrow, produced backwards into flattened spines; glabella smooth, cylindrical or ovate, about twice as long as wide, not reaching within about its own diameter of the front margin; one pair of small, basal, cephalothoracic furrows, or none. Eyes reniform, in the midst of the cheek (? smooth); eye-line unknown. Surface strongly granulated; one tubercle on each side of the anterior end of the glabella, the marginal row and a circle round each eye being larger than the rest. Body-segments unknown. Pygidium nearly resembling the cephalothorax in size and form, rather more pointed, strongly trilobed, and with a thickened prominent margin; axal lobe about as wide as the lateral lobes, of about seventeen narrow segments; lateral segments about seven, divided from their origin, each terminating in a large tubercle at the margin.

"The minute Trilobites for which I propose the present genus are very distinct in habit from those of other genera, and as two or three species are now known, it seems desirable to place them together under one name. They are the smallest perfect Trilobites known, from two to three lines being the greatest width they have been seen to attain. *Phillipsia Maccoyi* of Captain Portlock's 'Geol. Report on Londonderry,' &c., certainly belongs to this genus, and is at first sight difficult to distinguish specifically from the Australian species. The Irish species alluded to was collected by the writer from the Lower Carboniferous Limestone of Kildare, and sent to Captain Portlock for his 'Monograph of Irish Trilobites,' under the impression that it formed the type of a new genus and species, but probably from there being but one specimen it was placed provisionally by that author in his genus *Phillipsia*, from which it differs in its small, short glabella, smooth eyes, want of cephalothoracic furrows, &c. Having now examined numerous specimens of the Australian species, there can be no longer any doubt of the distinctness of the group from *Phillipsia* from the characters of the cephalothorax, and the pygidium is still more distinct. From those materials I have therefore drawn up the above characters, which it is believed will distinguish them easily from the other generic types. From the general similarity in the structure of the pygidium, I am inclined to refer the fossil which I have named *Phillipsia* (?) *discors* (' Synopsis of the Carb. Limestone Foss. of Ireland,' pl. 4, fig. 7, p. 161) to the same genus. This is also a very small Trilobite, the length of the pygidium being only three lines; and although referring it provisionally to *Phillipsia*, I suggested in the above work that it should, when better known, form the type of a distinct genus, which, however, it was not possible to frame until now."

23. Brachymetopus ouralicus, *De Vern*, sp. 1845. Pl. VIII, figs. 1—8.

Phillipsia Jonesii, *De Koninck*. Anim. Foss., p. 606, t. 53, fig. 6, 1844 (*non* Portlock).
— ouralica, *De Verneuil*. Geol. Russ., vol. ii, p. 378, tab. 27, figs. 16 *a*, *b*, 1845.
Brachymetopus ouralicus, *Morris*. Cat. Brit. Foss., p. 101, 1854.
— — *J. W. Salter & H. Woodw.* Cat. and Chart Foss. Crust., p. 16, fig. 118, 1865.
— — ? *V. von Möller*. Bull. Soc. Nat. Moscou, pp. 24—27, and pp. 67, 68, pl. ii, figs. 32—35, 1867.
— — *H. Woodw.* Cat. Brit. Foss. Crust., p. 28, 1877.
— — *H. Woodw.* Geol. Mag., Decade ii, vol. x, p. 534, pl. xiii, fig. 1, 1883.

Head-shield nearly twice as broad as it is long, slightly pointed in front; glabella small, tumid; very obtusely conical, only half the length of the head-shield, and one third its breadth, no short lateral furrows visible, only the two small basal lobes which truncate the posterior angles of the glabella; axal-furrow enclosing the glabella antero-laterally; neck-lobe narrow, rounded distinctly, separated by the neck-furrow from the glabella; eyes small, placed close to the glabella, prominent, reniform, surface smooth; no facial suture visible; free-cheeks convex confluent around the glabella, with a broad, flattened, slightly concave margin, the rim of which is slightly raised; posterior margin of free-cheeks separated by a furrow, continuous with the neck-furrow; the posterior angles of the head produced into short, slightly recurved spines; entire surface of head covered with small bead-like tubercles, five larger ones being placed around the front border of the glabella and one in advance of each eye on the free-cheeks.

Thorax not known.

Pygidium circular, one fourth wider than long, consisting of seventeen coalesced somites, the axis forming one third the entire breadth of the shield where it joins the thorax, but diminishing rapidly to a rather blunt extremity at about one fifth of its length from the margin of the tail-shield which encircles it; each ring of the axis ornamented by a row of small granular tubercles (somewhat irregular in size); about eight grooved pleuræ are seen on each side the axis, each pleura forming a raised rib, extending to and becoming wider at its rounded extremity near the border, marked by a row of small tubercles, and having a shorter intermediate rib, similarly ornamented, placed behind it in the furrow; the pleuræ, as well as the axis of the pygidium, are convex; the margin has a slight narrow rim around it.

Formation.—Carboniferous Limestone.

Locality.—Settle, Yorkshire; Caldbeck, Cumberland; Castleton, Derbyshire; Little Island and Blackrock, Cork; Ardshanbally, Limerick.

Many specimens of this species have been examined from the British Museum (Natural History), the Woodwardian Museum, Cambridge, the Museum of the Geological Survey of Ireland, and from the collections of the Rev. E. O. de la Hey, Cheshire, and Mr. Joseph Wright, F.G.S., of Belfast.

Subjoined is de Verneuil's original description of a pygidium of a Trilobite from the Carboniferous Limestone of Cosatchi-Datchi, East of the Urals ('Geol. Russia' (vol. ii, Palæontology, p. 378, tab. 27, fig. 16 *a, b*), and named by him *Phillipsia ouralica* (1845).

"Abdomen semielliptical in form and arched, median lobe produced, narrow, subtriangular, and composed of fourteen articulations of which only ten are visible, the four last being very small and compressed. The lateral lobes are larger than the median lobe, and surrounded by a straight granulated border. The nine pleuræ composing the tail are divided by a groove into two unequal parts, the narrower being ornamented by fine granulations, whilst the larger half bears tubercles twice as large. The median lobe is not prolonged to the posterior border, but is separated by a smooth and rounded border which divides the two series of lateral pleuræ.

"*Dimensions.*—Length 10 millimètres, breadth 12 mm. The breadth would be more considerable if the form of the abdomen was less arched. Median lobe 4 mm., lateral lobe 5 mm.

"*Affinities and differences.*—By the subdivisions of the lateral articulations, this species, of which the abdomen only is known, is easily recognised. It cannot be considered equivalent to *P. Jonesii*, of Portlock, although it has much resemblance. Nevertheless, if we consult De Koninck's figure of this species we see that it differs from it in the number of lateral articulations, which are from eleven to twelve, instead of nine. Secondly, by the mode of their double furrow which gives rise to ribs of unequal length, and also because these articulations are nearly united behind the median lobe, and not separated by a broad obtuse palette as in our species."

A pygidium of this species was first observed in 1844 by Prof. de Koninck in Belgium, but he erroneously referred it to *Ph. Jonesii*, of Portlock, which is a true *Phillipsia*, and has since been placed as a synonym under *P. derbiensis*. A second pygidium was next observed in the Carboniferous Limestone of Russia by M. de Verneuil, who named it, in 1845, *P. ouralica*, M'Coy, having in 1847 established the genus *Brachymetopus* for certain other closely similar forms. Prof. Morris, in 1854, placed both de Koninck's *P. Jonesii* and De Verneuil's *P. ouralica* together under *Brachymetopus*, giving de Verneuil's name the priority (the name *Jonesii* being a synonym of *P. derbiensis*, and therefore disqualified for use). I cannot ascertain certainly whether Sir R. I. Murchison published the 1854 edition of 'Siluria' *prior* to the second edition of Morris's Catalogue (1854) or not, but at p. 283 Murchison gives on his woodcut of Fossils (56) fig. 1, a head-shield of a Carboniferous Limestone Trilobite, named *Brachymetopus ouralicus*. Morris no doubt obtained his knowledge of the occurrence of this species in England from Salter, whom he quotes as his authority (at p. 101, op. cit.) C. L. Derbyshire (*Salter*). Murchison's woodcut, although very small, is, however, more like the head-shield of *B. Maccoyi*, having the cheek-spines more prolonged than is observed in *B. ouralicus*.

In Salter's and Woodward's 'Chart of Fossil Crustacea' (1865) Mr. Salter figured a head-shield under the name of *B. ouralicus*, in which no cheek-spines are shown at all. This was probably reproduced from a damaged specimen, as *B. ouralicus* always seems to have very short cheek-spines.

In 1867 Valerian von Möller figured a head of *Brachymetopus*, from the Ural Mountains, which he placed under *B. ouralicus* with a note of interrogation, and he makes the following observations upon it (op. cit., p. 27).

"*Brachymetopus uralicus* was first described by De Verneuil as *Phillipsia ouralica* in the year 1845, but De Verneuil knew only the tail-shield. A few years later a cephalothorax was found in the Carboniferous Limestone of Derbyshire, which showed by its head decoration that it was so nearly related to De Verneuil's species that English palæontologists did not hesitate to identify it with that species. On the other hand, the English specimen shows in the principal features a marked resemblance to *P. Maccoyi*, Portl., so that no doubt now remains as to its connection with *Brachymetopus*. Under this generic name the fossils of our own country were first made known in Morris's 'Catalogue of British Fossils,' and later in Murchison's 'Siluria' (both in 1854). The result of the last year's researches decides me still more in the conclusion that the specimen from Derbyshire quite differs from the typical Uralian form, and therefore belongs to some other species; my reasons for this I will find an opportunity to give more fully."

At p. 67 he observes: "The cephalothorax under consideration is distinguished from the other species of *Brachymetopus* by the rounded cheek-plates, the wide front, and by the smooth rounded border, and by a different distribution of the bead-like ornamentation of the surface. I believe this specimen belongs to the same species as the pygidium, which has been described as *Phillipsia ouralica* by De Verneuil; at all events, both offer a great similarity in the ornamentation of the surface. If time should prove that my idea is correct then the cephalothorax from the Carboniferous Limestone of Derbyshire, figured in Murchison's 'Siluria' as *Brachymetopus ouralicus,* belongs to a separate species. It differs from the Ural specimens in the pointed cheek-spines, the remarkably small glabella, and the conspicuously small eyes.

In the last-mentioned characters the cephalothorax from Derbyshire differs from all other *Brachymetopi*, and I think therefore that it belongs to the already mentioned tail called *Phillipsia Jonesii* of De Koninck. In Morris's 'Catalogue of British Fossils,' 1854, p. 101, we find these Belgian specimens united with *B. ouralicus*, but they differ from it in the club-shaped widened form of the tail-segments, and the irregularity of the arrangement of the tubercles upon their surface."

It will readily be seen how the difficulty which von Möller experienced arose, when we bear in mind the fact that de Verneuil's *Phillipsia* (*Br.*) *ouralica* was established upon a pygidium only; and de Koninck's *Phillipsia* (*Br.*) *Jonesii*, upon a similar tail from Belgium. M'Coy, who established the genus *Brachymetopus*, did so upon the head and tail of another species, and makes no mention of either the Russian or Belgian specimens.

I think we are justified in concluding (*a*) that the small head-shield figured by Salter in Murchison's 'Siluria' (1854) as *B. ouralicus* (p. 283, fig. 1, Fossils 56) is most probably *B. Maccoyi;* (*b*) that the figure in Salter's and Woodward's chart of Fossil Crustacea (fig. 118, 1865) destitute of cheek-spines is a badly drawn or imperfect shield of *B. ouralicus;* (*c*) that von Möller's figure of a head-shield of *B. ouralicus* (?) is either quite a distinct species, *without cheek-spines*, and more highly ornamented, or that the Russian artist in Moscow who prepared the plate may have embellished it a little more than the specimen perhaps warranted, in placing the large tubercle on the centre of the glabella, in omitting the basal lobes on the glabella, and giving the angles of the cheeks a rounded contour. This last-named feature may be due to the fact that the cheek-spines were not preserved; of course, there may have been two forms, one with cheek-spines, and one without.

I think for the present, however, we may safely retain the name *ouralicus* for those English specimens which agree in having a pygidium like that figured by de Verneuil associated with a head-shield like those on our Pl. VIII, figs. 1, 2, 3, 4, and 5. Should, however, fuller information be obtained rendering it necessary to separate the Russian species, these must of course retain the original name of *ouralicus*, and for this English form I would then propose that the name *ornatus* should be adopted.

24. BRACHYMETOPUS MACCOYI, *Portlock*, sp., 1843. Pl. VIII, figs. 9—13.

PHILLIPSIA MACCOYI, *Portlock*. Geol. Rep. Londond., p. 309, t. 11, fig. 6, 1843.
BRACHYMETOPUS MACCOYI, *M'Coy*. Ann. and Mag. Nat. Hist., vol. xx, p. 230, 1847.
— — *Morris*. Cat. Brit. Foss., p. 101, 1854.
— — *H. Woodw.* Cat. Brit. Foss. Crust., p. 27, 1877.
— — *H. Woodw.* Geol. Mag., Decade ii, vol. x, p. 535, pl. xiii, fig. 2, 1883.

Carapace not quite twice as broad as it is long, rather pointed in front; surface sparsely covered with small tubercles; base of the glabella equalling one third the breadth of carapace, and one half the length of shield (without the neck-lobe); basal lobes small, distinct; no facial suture visible; eyes large, placed on the highest point of cheeks, and wider apart than the base of the glabella; rim of the head-shield ornamented by a single row of tubercles, margin strongly grooved or channelled, outer rim slightly raised; posterior border of cheeks marked by a distinct margin equal in width to the neck-lobe, and separated by a groove corresponding to the neck-furrow; latero-posterior angles produced into spines; thoracic somites not known (probably nine in number). Axis of pygidium composed of fifteen coalesced somites, tapering to an obtuse extremity; each somite having about five small tubercles on the axis, and about as many on the eight simple lateral lobes; ribs ending abruptly near the margin of pygidium.

Formation.—Carboniferous Limestone.

Localities.—Ballysteen and Monaster, Ireland.

B. Maccoyi approaches very closely to the preceding species (*B. ouralicus*), but differs in its smaller size, in the more marked and raised rim of the anterior border to the head-shield, in the less highly ornamented surface of the carapace, in which the eyes are larger in proportion and placed wider apart, and the angles of the cheeks are produced into spines which probably reached to the sixth free segment of the thorax. The thorax is unknown. The pygidium is detached, but occurs in the same matrix and from the same locality as the head-shield, and there seems no reason to doubt their relation to each other. The side-ribs of the pygidium of *B. Maccoyi* differ from those of *B. ouralicus* in being simple, whereas in the latter they are grooved, and each appears like a double rib.

This species has only been met with in Ireland. Our figures are drawn from specimens in the collections of Joseph Wright, Esq., F.G.S., of Belfast, and the Geological Survey of Ireland, Dublin.

Subjoined we give Portlock's original description of *Phillipsia* (*Brachymetopus*) *Maccoyi*, Portlock, *sp.*, 1843.—" Of this minute but very beautiful species the cephalothorax alone has as yet been discovered. It is placed provisionally in this genus (*Phillipsia*), though, from its diminutive size some of the characters of the genus cannot with certainty be exhibited, such for example, as the furrows of the glabella or the reticulation of the eyes. Cephalothorax elevated; general form oval, approaching to semicircular; length ·17″, breadth ·25″; glabella very small, in length rather more than two thirds of the length, and in breadth one fourth of the total breadth of the cephalothorax; the eyes apparently smooth, lunate, and equal in length to two thirds of the height of the glabella; the cheeks are large, slightly convex, separated from the margin by a furrow which joins the neck furrow, and is continuous, passing in front round the glabella. Margin or wings (free-cheeks) elevated, and ending at the posterior angles in prolonged points; the whole surface is covered with granulations, which are arranged along the ridge of the margin and the supra-palpebral line of the eyes like beads." (' Geol. Londonderry,' p. 309.)

Prof. M'Coy, in his ' Synopsis of Carboniferous Fossils of Ireland,' p. 162, refers to this species as follows:

" The only specimen which has occurred of this species I collected myself from the Carboniferous Limestone of Kildare; it was lent to Captain Portlock for his ' Monograph of Irish Trilobites,' as I conceive it to be not only a new species, but the type of a new genus. I am still inclined to think it cannot be ranked with any known genus; the very small size of the glabella, and large size of the eyes, distinguish it from the most nearly allied, and from *Phillipsia*, with which Captain Portlock has ranked it; it is distinguished by the want of the cephalothoracic furrows. As I have, however, no means of examining the specimen now, I cannot characterise it. Length of cephalothorax two lines, width three lines."

25. Brachymetopus discors, *M'Coy*, sp., 1844. Pl. VIII, fig. 15.

Phillipsia (?) discors, *M'Coy*. Synop. Carb. Foss. Irel., p. 161, t. 4, fig. 7, 1844.
Brachymetopus discors, *M'Coy*. Ann. and Mag. Nat. Hist., vol. xx, p. 230, 1847.
— — *Morris*. Cat. Brit. Foss., p. 101, 1854.
— — *Salter & H. Woodw.* Cat. and Chart Foss. Crust., p. 16, fig. 120, 1865.
— — *H. Woodw.* Cat. Brit. Foss. Crust., p. 27, 1877.
— — *H. Woodw.* Geol. Mag., Decade ii, vol. x. p. 536, 1883.

This species is founded on a pygidium only, and was in the first instance (1844) referred by M'Coy to *Phillipsia*, with a note of interrogation. Later on, namely, in 1847, M'Coy proposed that this species should be placed under his genus *Brachymetopus*. We have no further information to give concerning *B. discors*, and the only additional specimen we have seen is from the Carboniferous Limestone of Little Island, Cork, and consists of an equally small pygidium to that described by M'Coy, and was obtained by Joseph Wright, Esq., F.G.S., of Belfast, to whom we are indebted for the opportunity of figuring the same.

The specimen is very obscure and not well preserved.

The following is Prof. M'Coy's original description of *Phillipsia* (*?*) (now *Brachymetopus*) *discors:*

"*Specific Characters.*—Pygidium semielliptical; axal lobe reaching to the margin, one third less in width than the lateral lobes, very convex, composed of seventeen narrow segments, the third and fourth unite in the middle of the lobe to form one large tubercle, and towards the apex there are four or five small tubercles, irregularly disposed; the lateral lobes have only six large rounded segments, each terminating at the margin in a large rounded tubercle, and having usually between the margin and the axal lobe two other large, obtuse tubercles, one of these, on the third and last, being largest, and probably spiniferous; besides these there are a few irregular granules, especially towards the apex; all the lateral segments seem forked from nearly their origin."

Prof. M'Coy adds, "I have included this very remarkable Trilobite in the genus *Phillipsia*, Portlock, although I think there can be little doubt, if better known, it would form a genus distinct from any of those already constituted; I have named it from the great difference in number of the segments of the axal and lateral lobes of the pygidium. Length of pygidium three lines, width four and a half lines; width of axal lobe one line."

Formation.—Carboniferous Limestone.

Localities.—Millicent Clane, Kildare; Little Island, Cork.

26. Brachymetopus hibernicus, *H. Woodw.*, 1883. Pl. VIII, fig. 16.

Brachymetopus hibernicus, *H. Woodw.* Geol. Mag., Decade ii, vol. x, p. 536, pl. xiii, fig. 3, 1883.

Cephalothorax unknown. Pygidium broadly semicircular, 13 mm. wide and 8 mm. long; axis narrow, 3½ mm. broad and 6½ mm. long, composed of 11 coalesced segments, each alternate ring ornamented by a small tubercle on the centre; border composed of 10 rounded pleuræ, which extend to the margin, becoming gradually more and more oblique until the last pair become nearly parallel behind, uniting the extremity of the axis with the posterior border. There is no ornament on the pleuræ.

This detached pygidium slightly resembles *B. discors* in general form, but is quite distinct and well marked. I feel satisfied that it belongs to *Brachymetopus*, but it cannot be referred to any species hitherto described, nor have I seen any other but the one figured, which is a very perfect and well-preserved specimen.

Formation.—Carboniferous Limestone.

Locality.—Kildare, Ireland.

Obtained by the present Earl of Enniskillen (when Lord Cole), and now preserved in the British Museum (Natural History).

Genus 4.—Proetus, *Steininger*, 1830.

When the earlier pages of this Monograph were passing through the press, I was not aware of the existence of any evidence in favour of the introduction of the genus *Proetus* into the catalogue of our British Carboniferous-Limestone Trilobites.

So long ago, however, as 1861, Prof. Dr. James Hall had identified a Trilobite from the Waverley Sandstone of New York State as *Proetus auriculatus* ('15th Report on State Cabinet of New York,' 1862, p. 107).

Dr. R. Richter, in his paper "Der Kulm in Thuringen" (in the 'Zeitschrift der Deutschen geolog. Gesell.,' 1864, Bd. xvi, p. 160, Taf. iii, fig. 1), describes and figures a small Trilobite under the name of *Proetus posthumus*, but there does not appear any special reason, save the narrowing of the glabella in front, for separating it from the genus *Phillipsia*.

In 1865, Messrs. Meek and Worthen also described a Trilobite from the Lower Carboniferous series (Kinderhook Group), Jersey, Co. Illinois, which they named *Proetus ellipticus* ('Geol. Survey of Illinois,' vol. iii, "Palæontology," p. 460, pl. xiv, fig. 8).

A third specimen described by Prof. Meek as *Phillipsia Lodiensis* ('Palæontology of Ohio,' vol. ii) should, he thinks (op. cit. p. 324), be called *Proetus Lodiensis* (the figure, however, is unsatisfactory).

There seems, therefore, no reason to doubt that the genus *Proetus*, hitherto known here only in the Silurian and Devonian, may have extended upwards into our Lower-Carboniferous series, as we find it on the American Continent.

It is important also that we should bear in mind the close family relationship which exists between *Phillipsia*, *Griffithides*, and *Proetus ; Brachymetopus* alone enjoying any strongly-marked distinctive generic peculiarities.

The general form of the body is oval; and the trilobation very distinct through the entire length of body. The head is less than a third of the total length; the pygidium is rather longer than the head; the head-shield is always surrounded by a border, consisting of an exterior raised rim and an inner groove or furrow; the border is sometimes prolonged into a spine at the angle of the free-cheeks. The posterior margin of the head is formed by the grooved and furrowed border of the free-cheeks on each side and by the two basal lobes and the neck-lobe, which are separated from the glabella by a very distinct and deep furrow; the neck-lobe is broader than the free thoracic somites which follow it; the glabella is usually rounded and gibbous in front, but does not overhang its anterior border. (Barrande states that there are three pairs of short lateral grooves on its surface, although not always to be distinguished.) The axal furrows which surround the glabella are very distinct; the facial suture (which divides the fixed-cheek from the free-cheek) crosses the frontal border just in a line with the compound eye, above which it expands, forming a rounded palpebral lobe; then, passing down close to the line of the axal furrow, it diverges outwards and crosses the posterior border obliquely behind the line of the orbit. The free-cheek is triangular, its surface is convex, and upon the highest point is placed the large compound reniform eye, which either sometimes exhibits a faceted surface or is quite smooth, according to the state of its conservation.

Free thoracic segments, varying from eight to ten (*Proetus Barrandii*, Roemer, Devonian, Harz, has eight somites; nine and ten is the common number for the Silurian species). The axis is always strongly arched, and does not exceed the pleuræ in breadth; the breadth of the axis diminishes very gradually to the posterior extremity; the pleuræ are more or less bent at the fulcral point, and have their extremities either pointed or rounded, and their anterior margin faceted for rolling up.

The pygidium varies in its elevation, but the axis is always raised above the margin, and diminishes to a blunt extremity, leaving a smooth border beyond; the number of coalesced segments in the tail-shield varies (Barrande says from four to thirteen in species found in Bohemia); the pleuræ do not extend to the margin of the pygidium, which is often smooth.

The surface of the test is most frequently smooth or finely granulated; in a few species it is striated; and rarely it presents a combination of both kinds of ornamentation.

Mr. J. W. Salter, in his "Palæontological Appendix" to the 'Memoirs of the Geological Survey,' 1858, vol. ii, part i, p. 337, writes as follows concerning the generic characters of *Proetus :*

"A very usual character of this genus is the possession of a strong tubercle, terminating the neck-segment on each side, and nearly separating it. Burmeister, however, in his second edition, has considered the species having this thickening and the obscure glabella furrows more strongly marked as forming a distinct genus, which he calls *Æonia*. M'Coy had anticipated him by a few months in the name of *Forbesia* without referring to *Proetus*. We prefer, with Lovèn, to consider both as belonging to one genus.

"The glabella shows two or three lateral furrows, but is sometimes quite smooth. Body-rings ten. The tail has but few lateral furrows, seldom more than seven or eight, and in this, as well as in the additional body-segment, the Silurian genus differs from the Carboniferous *Phillipsia* and *Griffithides*, which are else very nearly related to it."

27. PROETUS? LEVIS, *H. Woodw.*, 1883.[1] (See Woodcut, fig. 1.)

PHILLIPSIA BRONGNIARTI ?, *Baily*. Mem. Geol. Surv. Ireland, Expl. Mem. Sheets 102 and 112, 2nd edit., p. 19, 1875.

PROETUS ? LEVIS, *H. Woodw*. Geol. Mag., Decade ii, vol. x, p. 446, and woodcut, 1883.

Cephalothorax unknown.

Pygidium—22 mm. broad and 16 mm. long; smooth, semicircular, one fourth broader than long; axis convex, 7 mm. wide, one third the breadth of the pygidium at its anterior border, smooth, moderately elevated; axal furrows broad; where decorticated showing evidence of the coalescence of about twelve somites or rings 12 mm. long; tapering to a blunt extremity, which does not reach to the posterior margin, but leaves a smooth border 4 mm. wide behind it; pleuræ slightly convex, very indistinctly furrowed; margin

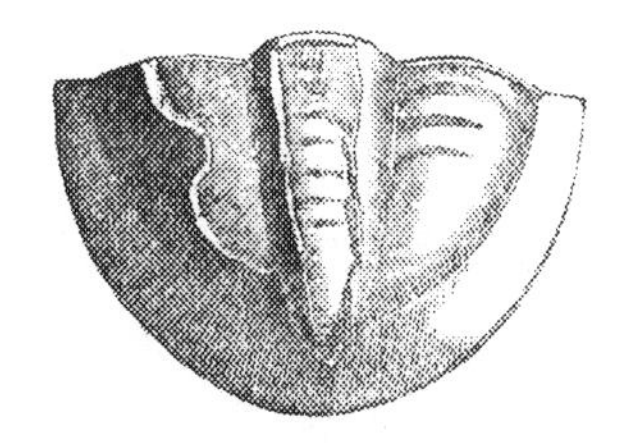

FIG. 1. Pygidium of *Proetus ? levis*, H. Woodw. Carboniferous Limestone, Moneenalion, Co. Dublin. Enlarged nearly twice natural size.

[1] The existence of this specimen was not known until too late for insertion in the earlier pages of this Monograph. It should naturally be placed as the first genus instead of the fourth in our series.

entire, smooth, broad; general surface smooth, and destitute of ornamentation of any kind save a simple rib-like furrow and ridge where the pygidium unites with the free thoracic segments.

Formation.—Carboniferous Limestone.

Locality.—Moneenalion Commons, Co. Dublin. From the Survey Museum, Dublin. Kindly lent by Prof. E. Hull, M.A., LL.D., F.R.S., Director.

The specimen here described is the same recorded by Mr. W. H. Baily, F.L.S., in the Explanatory Memoir to Sheets 102, 112 (p. 19), as "*Phillipsia Brongniarti*," and was obtained from the Upper Limestones on the south side of Dublin, at Moneenalion Commons, about one mile south-east of Castle Bagot (Sheet 111). In the Explanation of Sheet No. 111 Mr. G. V. Du Noyer wrote (p. 21): "The general aspect of the Limestone varies between that of a palish and a dark grey compact rock; it is usually very fetid, and contains layers of chert.

"Some beds are very fossiliferous, and the Trilobite (*Griffithides*) is not uncommon in them; others consist almost entirely of crinoid fragments, large *Productæ* occurring sometimes in layers."

The specimen under consideration is embedded in dark (almost black) fetid crystalline limestone, full of crinoidal fragments and of Brachiopoda.

A careful comparison of this specimen with the pygidium described by Phillips (formerly called "*Ph. Brongniarti*," but now placed in the genus *Griffithides* under Phillips's original name of *obsoletus*) has satisfied me that they cannot possibly be placed together.

"*Ph. Brongniarti*" = *G. obsoletus*	compared with *Proetus levis*.
(*a*) Axis of pygidium nearly equal to half the breadth.	(*a*) Axis less than one third the entire breadth.
(*b*) Axis extremely gibbous.	(*b*) Axis but little elevated.
(*c*) Pleuræ very convex.	(*c*) Pleuræ almost flat.
(*d*) Axis and pleuræ very distinctly annulated.	(*d*) Axis and pleuræ very nearly smooth.
(*e*) Pygidium one fourth broader than long.	(*e*) Pygidium one third broader than long.

The fact of this abdominal shield from Co. Dublin being so unlike any one belonging to *Phillipsia* or *Griffithides* led me to compare it with those of other genera, and I was at once struck with its resemblance to the pygidia of *Proetus*. Seeing that in 1861, Hall had applied this generic appellation to a Carboniferous Trilobite from America, there cannot, I think, be any very great objection to its use here.

The pygidium is certainly new, and agrees with that of *Proetus* (*e.g. P. latifrons*) better than with any other genus with which I am acquainted.

Notes on the Discovery of Trilobites in the Culm-Shales of South-East Devonshire.

Although the Culm-deposits of Devonshire have long been known and studied, it has been a matter of considerable doubt as to the exact horizon which they really occupy. The Geological Surveyors have, it is true, spent much time in re-examining certain parts of the county, but owing to the smallness of the scale of the Ordnance Survey map (only one inch to the mile), and the inaccuracy of the topography, but little could be done to show the detailed work resulting from these later investigations.

Prof. J. Beete Jukes, F.R.S. ('Quart. Journ. Geol. Soc.,' 1866, pp. 320—371), and subsequently Mr. R. Etheridge, F.R.S. (op. cit., 1867, pp. 568—698), described the whole of the northern portion of the county afresh, whilst Mr. Horace B. Woodward, F.G.S., Mr. Clement Reid, F.G.S., Mr. W. A. E. Ussher, F.G.S., as representing the Geological Survey, have been engaged upon the more southern parts. Added to this Dr. Harvey B. Holl, F.G.S., Mr. A. Champernowne, F.G.S., and Mr. Jno. E. Lee, F.G.S., have contributed not a little to the elucidation of difficult parts of the Geology of South-Eastern Devon, whilst Mr. Townshend Hall, F.G.S., has done equally useful work in the northern area.

1839.—Sir H. T. de la Beche[1] notices the Culm-formation, and intimates (p. 117) that Prof. J. Phillips regarded the shale-fossils as belonging to the Carboniferous Limestone.[2] The list of plants which he gives contains a mixture of species, but many of those quoted are probably true Coal-Measure plants which do not occur in the Culm proper.[3]

1840.—Prof. Sedgwick and Sir R. I. Murchison in their memoir "On the

[1] 'Report on the Geology of Cornwall, Devon, and West Somerset,' 1839. See also 'Trans. Geol. Soc.,' 2nd series, vol. iii, p. 163.

[2] Judging from the *apparent* dip of the Culm-Measures beneath the Devonian rocks at Chudleigh, Sir H. de la Beche considered that this Limestone (which is now proved by its fossils to be the *Goniatites intumescens* Stage of the Upper Devonian, Roemer, 'Geol. Mag.,' 1880, pp. 145—147) was included in the Carbonaceous series, and as such it was originally engraved and published in his sections and coloured in the Geological Survey Map (Clement Reid, 'Geol. Mag.,' 1877, p. 454).

[3] Mr. R. Kidston, F.G.S., mentions (in a letter to the author) the following plants as determined by him from the Culm, viz.:

Asterocalamites scrobiculatus, Schlot., sp. (= *Bornia radiata*, Brong.).
Calamites Roemeri, Göpp.
Sphenopteris, sp. nov.
Lepidodendron Rhodeanum (?), Sternb.
Lepidophloios, sp.
Halonia (fruiting branch of *Lepidophloios*).
Sigillaria (?).
Stigmaria ficoides, Brongt.

"All these plants," he adds, "have a 'Calciferous Sandstone' facies, and are equivalent to the 'Culm' of Germany."

Physical Structure of Devonshire, and on the Sub-divisions and Geological Relations of the Older Stratified Deposits,"[1] devote pp. 669—684 to a consideration of the "Culmiferous Series, its Relation to the other Formations, Structure, and Fossils." They mention (p. 678) that "in Ugbrook Park, near Chudleigh [in close proximity to Waddon-Barton, where the Trilobites were discovered by Mr. Lee], there is a large development of Culm Sandstone as coarse as Mill-stone grit, and passing into a conglomerate form; over it are some beds of more thin-bedded grey sandstone, not to be distinguished from a Coal-Measure sandstone, and containing very fine vegetable impressions, among which are well-marked *Calamites*. Indeed, through the whole of the upper group we are describing, vegetable impressions, though rarely so perfect as to give anything like specific characters, are extremely abundant." They add, "All the beds are intersected by numerous open joints, which in the coarser contorted beds are very irregular in their directions. But when the beds have a finer flaggy or shaly structure, the joints often become parallel (especially in a direction nearly transverse to the strike) so as to separate the strata into prismatic masses"[2] (p. 679).

"Among the more calcareous bands some are fossiliferous, containing a great abundance of at least two genera of bivalve shells; one a *Posidonia*" (*Posidonomya Becheri*, Bronn) the other of a genus not ascertained, but regarded as a marine shell. "In the same part of the series are *Goniatites* of at least two species, both of which are unquestionably marine, and (according to Professor Phillips) identical with *Goniatites* of the Yorkshire Coal-field."

After quoting Prof. Lindley's determinations of the Plants (pp. 681—682), the authors conclude: "On the whole, considering that the culmiferous rocks of Devon form a distinct group, with a peculiar mineral type (unlike the older groups, but nearly resembling the culmiferous beds of Pembrokeshire)—that they overlie all the other groups, and are overlaid by no rock newer than the New Red Sandstone—that, notwithstanding the paucity of fossils in the black limestone (in which respect it resembles the 'calp' of Ireland), there are in it one or two species not separable from known Mountain Limestone fossils, and, finally, that the flora of the Upper Culms, as far as it has been ascertained, agrees specifically with the known flora of the Carboniferous period; we think we have strong direct evidence to establish our position, "that the Upper Culm Strata of Devon are the geological equivalent of the ordinary British Coal-fields."

1842.—Mr. R. A. C. Austen,[3] whose paper was *read* December 13th, 1837, describes the culmiferous deposits of the South-East of Devonshire, and particularly

[1] 'Trans. Geol. Soc. Lond.,' second series, vol. v, 1840 (*read* June 14th, 1837).

[2] This paragraph gives a very exact description of the lithological characters of the beds at Waddon-Barton by Chudleigh, containing the Culm Trilobites.

[3] 'Trans. Geol. Soc. Lond.,' 1842, 4to, second series, vol. vi, "On the Geology of South-East Devonshire."

at Ugbrook Park, near Chudleigh, and other adjacent places, and states that Prof. Sedgwick considered them as a portion of the culmiferous beds of the centre of the county (p. 457).

Mr. Austen quotes a list of the plants, and adds (pp. 461-2), "This Flora, so far as it goes, is that of the Carboniferous period. In the black limestones occur *Goniatites mixolobus*, Phil., and *Goniatites crenistria*, Phil., Mountain Limestone species."

1867.—Sir Roderick I. Murchison, in the 1867 edition of 'Siluria' (p. 273), writes:

"Now, although this over-lying series is in mineral aspect as much unlike the Carboniferous strata of most other parts of Britain as the rocks of N. Devon are unlike the ordinary Old Red Sandstone of England and Scotland, we have proofs of fossils, besides the analogy with Pembrokeshire before spoken of, that the black limestones of Swimbridge and Venn, &c., with *Posidonomyæ*, do represent, on a miniature scale, a part of the Mountain or Carboniferous Limestone, that the next series of white grit and sandstone of Coddon Hill, &c., stands in the place of the Millstone-Grit, and that the overlying courses of Culm, with many remains of Plants, are consequently the equivalents of some of the lower coal-bearing strata of other tracts. In short, no one denies that in the Culm series of Devonshire we have the representatives of the Lower Carboniferous Strata."

1868.—Dr. Harvey B. Holl, in his paper "On the Older Rocks of South Devon and East Cornwall,"[1] describes the Carbonaceous Rocks or Culm-Measures very fully. He mentions the hard slates at Waddon-Barton over-lying the limestone, full of *Goniatites* and *Posidonomyæ*, above which are the typical Carbonaceous Sandstones quarried at Ugbrook Park. In conclusion, he refers to the memoir by Sedgwick and Murchison, and adds, "It is to these authors that we are indebted for having first pointed out the true position of these (Culm) rocks in the geological scale, when, by means of the included plant and other fossil remains, they identified them with the Coal-Measures of South Wales."

1875.—Mr. Townshend M. Hall,[2] in his 'Notes on the Anthracite Beds of North Devon,' writes, "In the North-Devon district the anthracite (Culm) is found in the Millstone-Grit, a series of beds belonging to the Carboniferous formation, but of an age immediately antecedent to that of the true Coal-Measures." The list of Culm-plants given by Mr. Townshend Hall, however, needs revision.

1876.—Mr. Horace B. Woodward, in his 'Geology of England and Wales,' pp. 106—111, gives a concise account of the literature of the Devonshire Culm-Measures. "Looked at in a large way, they consist of a series of shales, grits, chert-beds, with beds of limestone here and there." "Some authorities have placed them, generally, on the horizon of the Millstone-Grit, but there seems

[1] 'Quart. Journn. Geol. Soc. Lond.,' 1868, vol. xxiv, p. 401.

[2] 'Trans. Devonshire Association,' 1875.

reason to include with them representatives of at least a portion of the true Coal-Measures, and possibly also of the Carboniferous Limestone."

(1) The NORTH DEVON series of Carboniferous deposits about Barnstaple and Bampton is thus given by Prof. Phillips (p. 189):[1]

(*a*) "Upper part anthracitiferous, and containing ironstone, and by these characters agreeing with the Coal-deposits of Pembrokeshire. This is in general a Gritstone series, with plants of the Coal-formation."

(*b*) "Coddon Hill cherts, black-grits, jasper-rock, lydian-stones and shales of considerable, but variable, thickness; 1500 to 2000 feet (according to the Rev. D. Williams)."

(*c*) "Limestone and black shale with *Posidonomya*, *Goniatites*, &c. = *Posidonomya* (*Posidonia*) limestone of Swimbridge and Venn."

(*d*) "Black shale group."

(2) The SOUTH DEVON strata about Trescott and Lew Trenchard have been thus divided (op. cit., p. 194):

(*a*) "Gritstone group of Central Devon."

(*b*) "Upper shale group—dark shales, carbonaceous grits and shales (equal to the Coddon Hill series."

(*c*) "Calcareous group—limestone of dark colour, and irregular bedding, with shales (*Posidonomya*)."

(*d*) "Lower shale group, with few fossils (no slaty cleavage)."

1882.—Dr. A. Geikie, F.R.S. (the present Director-General of the Geological Survey of Great Britain), writes in his 'Text-Book of Geology' (p. 748) as follows:

"In Moravia, Silesia, Poland, and Russia, the Carboniferous Limestone reappears as the base of the Carboniferous system, but not in the massive calcareous development which it presents in Belgium and England. One of its most characteristic phases is that to which the name 'Culm' (applied originally to the inferior slaty coal of Devonshire) has been given, when it becomes a series of shales, sandstones, greywackes, and conglomerates in which the abundant fauna of the limestone is reduced to a few molluscs (*Productus antiquus*, *P. latissimus*, *P. semi-reticulatus*, *Posidonomya Becheri*, *Goniatites sphæricus*, *Orthoceras striatulum*,[2] &c.). The *Posidonomya* particularly characterises certain dark shales known as 'Posidonomya schists.' About fifty species of plants have been obtained from the Culm, typical species being *Calamites transitionis*, *Lepidodendron veltheimianum*,

[1] 'Figures and Descriptions of the Palæozoic Fossils of Cornwall, Devon, and West Somerset,' by Prof. John Phillips, F.R.S., 1841. 8vo.

[2] = *O. striolatum*, Sandb. The above-mentioned shells, which are all marine, occur in the Calciferous Sandstone around Edinburgh and in Fifeshire (see paper by Mr. R. Etheridge, junr., "On the Invertebrate Fauna of the Lower Carboniferous or Calciferous Sandstone of Edinburgh, &c.," Quart. Journ. Geol. Soc., 1878, vol. xxxiv, pp. 1—26, plates i and ii).

Stigmaria ficoides, Sphenopteris distans, Cyclopteris tenuifolia. This flora bears a strong resemblance to that of the Calciferous Sandstones of Scotland."

We are indebted to Mr. John Edward Lee, F.S.A., F.G.S., for the discovery of Trilobites in the lower Culm-shales of Waddon-Barton about two years since. The spot where the discovery was made, and which, up to the present time, is the only locality that has yielded these organisms, is the bankside of a steep lane leading at right angles from the ridge-road between Chudleigh and Haldon, and near the village of Waddon-Barton.

These Goniatite-shales, which break up (as so well described by Sedgwick and Murchison) into small cuboidal or prismatic fragments, are full of minute marine organisms. A list of these had been prepared by Mr. Lee, and to this Mr. Robert Etheridge, junr., and myself have contributed some additional species.

The main interest consists in the fact that these Goniatite-bearing shales agree almost exactly with the beds of corresponding age recently most carefully worked out and described by Prof. Dr. A. von Koenen, late of Marburg (now of the University of Göttingen), in a paper entitled " Die Kulm-Fauna von Herborn."[1]

Prof. Dr. A. von Koenen's paper is accompanied by two plates of Culm fossils, but he does not give figures of the two species of Trilobites which he refers to the genus *Phillipsia*, and he adopts the specific determinations of von Meyer and of Sandberger, whose figures are, however, not very satisfactory. We shall refer to these Culm species again further on.

The following is a list of species obtained by Prof. Dr. A. von Koenen from the Culm of Herborn, near Dillenberg :[2]

1. *Phillipsia æqualis*, v. Meyer.
2. *P. latispinosa*, Sdbg.
3. **Cypridina subglobularis*, Sdbg.
4. **Goniatites mixolobus*, Phill.
5. *G. crenistria*, Phill.[3]
6. *Aptychus carbonarius*, v. Koen.
7. *Orthoceras scalare*, Gldf.
8. **O. striolatum*, v. Meyer.
9. *O. cf. giganteum*, Roemer.
10. *O. cf. inæquale*, Roemer.
11. *O. undatum*, Flem.
12. *Orthoceras*, sp.
13. *Bactrites*, sp.
14. *Gyroceras serratum*, de Kon.
15. *Nautilus*, sp.
16. *Nautilus*, sp.
17. *Hyolithes Roemeri*, v. Koen.
18. *Terebratula hastata*, Sow.
19. *Camarophoria papyracea*, Roem., sp.
20. *C. triplicata*, v. Koen.
21. *Spirifer ? makrogaster*, Roemer.
22. *Orthis concentrica*, v. Koen.
23. *Productus cf. sublævis*, de Kon.
24. **Chonetes deflexa*, v. Koen.
25. **C. rectispina*, v. Koen.
26. *Pecten densistria*, Sdbg.[3]

[1] See Leonhard und Geinitz's ' Jahrbuch für Mineralogie, &c.,' 1879, pp. 309—346, pls. 6 and 7.

[2] Those marked with a * are considered to occur in the Culm-shales of Waddon-Barton, Devon. Probably several others should also be identified, but they require actual comparison with each other to determine this with certainty.

[3] Mr. Jno. E. Lee believes these should also be added to the Culm-list of fossils from Devonshire.

27. *P. Losseni*, v. Koen.
28. *P. prætenuis*, v. Koen.
29. *P. perovalis*, v. Koen.
30. *Aviculopecten cf. papyraceus*, Sow.
31. **Avicula lepida*, Gldf.
32. *A. latesulcata*, v. Koen.
33. *A. Kochi*, v. Koen.
34. **Posidonomya Becheri*, Gldf.
35. *Myalina mytiloïdes*, v. Koen.
36. *Arca Rittershauseni*, v. Koen.
37. *A. cf. arguta*, Phill.
38. *A. Decheni*, v. Koen.
39. *Poteriocrinus regularis*, H. v. Meyer.
40. *Lophocrinus speciosus*, H. v. Meyer.
41. *Cyathophyllum*, sp.
42. *Listrakanthus Beyrichi*, v. Koen.
43. *Cladodus striatus*, Ag. ?
44. Fish-jaw. ?

List of fossils from the Lower Culm-shale of Waddon-Barton, near Chudleigh, Devonshire, revised and augmented by Mr. R. Etheridge, junr., and Dr. H. Woodward, from Mr. Lee's collection, and from other sources.

Genera, Species, and Remarks.

1. *Orthoceras striolatum*, Sandb. (chiefly as external casts).
2. — sp. (there are probably more than two species of *Orthoceras*).
3. *Goniatites mixolobus*, Phill. (as figured by Roemer).
4. — *sphæricus*, Martin, sp. (as figured by Roemer).
5. *Posidonomya Becheri*, Bronn.
6. — *corrugata*, Eth. (? or young of *P. Becheri*).
7. *Pecten*, sp. nov. ? (of a Carboniferous facies, but differing from any figured by von Koenen).
8. *Pteronites*, sp. (form related to *P. persulcatus*, M'Coy).
9. — sp. (form related to *P. latus*, M'Coy).
10. *Avicula lepida*, Goldf.
11. *Chonetes rectispina*, von Koenen.
12. — *deflexa*, von Koenen.
13. *Spirifera Urii*, Fleming.
14. *Fenestella*, sp. (in the condition known as *Hemitrypa Hibernica*, M'Coy).
15. *Phillipsia Leei*, H. Woodw. (Pl. X, figs. 1, 2, 3, 4).
16. — *minor*, H. Woodw. (Pl. X, figs. 5, 6 *a*, *b*, 7, and 8 *a*).
17. — *Cliffordi*, H. Woodw. (Pl. X, figs. 8 *b*, 9, 10, 11, 12).
18. — *articulosa*, H. Woodw. (Pl. X, figs. 6 *c*, *d*, and 13).
19. ? *Bernix Tatei*, Jones (this form also occurs in the "Tuedian").
20. Casts of small corals (probably *Monticuliporidæ*).
21. Casts of small bodies (probably Sponge-Spicula).

It is highly probable, when more of the shale shall have been carefully examined, that many other small organisms will be added to our list, but the intractable nature of the matrix has precluded our doing more at present.

It may be interesting to record the fact here that in the "Tuedian" group or

Lower Carboniferous of Budle, Northumberland, Prof. G. A. Lebour has obtained *Posidonomya Becheri* identical with that of Devonshire.

"The Tuedian group (says H. B. Woodward) and the Lower Limestone shale are homotaxial with the Calciferous Sandstone group of Scotland" ('England and Wales,' p. 78).

Whatever may be finally decided to be the exact horizon of the Culm-Measures near Bideford I think it can no longer be denied that the *Posidonomya* and Goniatite shales of both North and South Devon are really (as suggested by Dr. A. Geikie, and now shown from their fossil contents by Mr. Jno. E. Lee) at the very base of the Carboniferous series, and are equivalent to the Lower Carboniferous series of the Rhenish Provinces and the Hartz.

There is little doubt also that the plant-remains which occur in the associated sandstones of the same regions are older than those of the Millstone-Grit series, and must be correlated with those of the Calciferous sandstones around Edinburgh.

The Trilobite remains from the Culm-shales of Waddon-Barton, Devonshire, are met with in the same condition as the *Goniatites* and other fossils with which they are associated. They are all highly compressed and often considerably affected by cleavage, causing them to be more or less distorted.

I recently visited Waddon-Barton, Chudleigh, and many of the localities for Culm-shale fossils with Mr. J. E. Lee, and I also spent a week in breaking up hundreds of pieces of the shale, two cartloads of which had been procured by Mr. Lee from Waddon-Barton with the permission of Lord Clifford. Out of this I obtained a large number of these Trilobites and other organisms with my own hands in addition to those already obtained by Mr. Lee.

Out of a series of nearly fifty specimens thus obtained, I have been able to determine four distinct species. They are all in a very fragmentary condition, the individuals varying from 10 millimètres in length (Pl. X, fig. 7 *a*) to 23 mm. and upwards (Pl. X, fig. 2).

As is the case in other deposits of Carboniferous age, it is most rare to meet with specimens having the head, thorax, and abdomen united. Only two approaching this state have been discovered as yet; the majority disclose evidence of detached pygidia, whilst head-shields and thoracic rings are but rarely found.

Making allowance for the effects of compression and distortion which the specimens have undergone, they are probably all referable to the genus *Phillipsia*, and strongly resemble in their mode of preservation the specimens of *Phillipsia Colei* from Ballintra and Carrickbreeny, Donegal, and of *Phillipsia truncatula* from Hook Point, Co. Wexford, Ireland.

28. Phillipsia Leei, *H. Woodw.*, sp. nov. Pl. X, figs. 1, 2, 3, and 4.

This is one of the largest of the Culm Trilobites, and is represented by numerous specimens.

The head-shield is semicircular in outline, the glabella occupying about one-third of its breadth at the widest part of the head, the glabella is moderately elevated, and is surrounded by the flattened border of the fixed cheek, which expands in front, forming a flat and somewhat broad semicircular border around the anterior portion of the glabella. Two small basal lobes are seen one on either side near the posterior margin of the glabella, and two short oblique furrows mark its sides. The neck-lobe is well defined and somewhat strongly arched, and is widest in the centre; the facial suture separating the free-cheek crosses the neck-lobe obliquely along its pleural portion close to the axal furrow; a deep furrow circumscribes the border and separates the raised inner portion of the free-cheek from the flattened margin of the shield which in its decorticated condition is seen to be ornamented by parallel lines. The angles of the head-shield are produced into strong rather broad spines about two-thirds as long as the head. The eyes are very small, semilunar, and often quite obliterated by compression.

Thoracic segments.—The axis of the thorax is somewhat wider than its pleuræ, the separate segments are distinctly marked by deep furrows, each of the pleuræ being marked by a central groove; their extremities are rounded; the thoracic segments were probably nine in number, but the whole number cannot be seen in any one specimen.

The pygidium is one-fifth broader than long, the axis forms one-third of its breadth at the proximal border, but diminishes very rapidly, terminating in a somewhat blunt point near the posterior margin.

There appear to be about fourteen coalesced segments in the axis of the pygidium represented by about nine grooved pleuræ on each side, surrounded by a narrow smooth border.

This species, which we have dedicated to the discoverer, Mr. John Edward Lee, F.S.A., F.G.S., of Villa Syracusa, Torquay, presents affinities with *Ph. gemmulifera*, *Ph. truncatula* and *Ph. Eichwaldi*, in all these species the angles of the head-shields are produced into lateral spines, and the flattened border of the glabella encircles the raised central portion, but the eyes in *Ph. Leei* are exceedingly small, whereas in all the other species of *Phillipsia* they are very large and prominent.

Ph. Leei differs from *Ph. Colei* in possessing cheek-spines. The pygidium of *Ph. Leei* is also distinct, being narrower and more pointed in its axis; the tail-shield itself is also more triangular in outline.

We have compared *Phillipsia Leei* with *Proetus posthumus* of Richter[1] to

[1] 'Zeitsch. Deutsch. geolog. Gesells.,' 1864, Bd. xvi, p. 160, Taf. iii, fig. 1.

which it approaches, but in Richter's figure the glabella is narrow in front and broader behind, whereas our Culm form is just the reverse. Prof. A. von Koenen, p. 312, op. cit., places Richter's *P. posthumus* with *Phillipsia æqualis*, H. von Meyer, and observes, "As von Meyer expressly says that the glabella is reduced in front, there is no doubt that Burmeister was in error in figuring as this species a form with a club-shaped glabella. Emmrich's restored figure does not give a good representation of the species; the head-shield is too long, the eyes placed too far in front, and the glabella too slightly reduced in front. The form placed by Roemer under *Phillipsia latispinosa*, from the Silesian Culm of Bantsch appears to me, on account of the very wide glabella, to belong to *P. æqualis*. Near Nehden I have found an example of which the tail is 15 mm. wide and 10 mm. long, and fragments of still larger dimensions occur at Aprath" (op. cit., p. 313).

We figure here a specimen of Culm with two compressed Trilobites thereon referred to *Phillipsia*, which were obtained at Aprath by Mr. J. E. Lee, F.G.S., of Torquay; but the cheek-spines are wanting in both, although the eyes can be discerned with a high power, and there are faint traces on one of the obliquely transverse furrows on the glabella. A reference to *Cylindraspis latispinosa*, of Sandberger (taf. iii, fig. 4 and 4 *a*) shows that the glabella of this species is also more pointed in front than in our British species; the facial suture is close to the glabella, as in other *Phillipsiæ*.

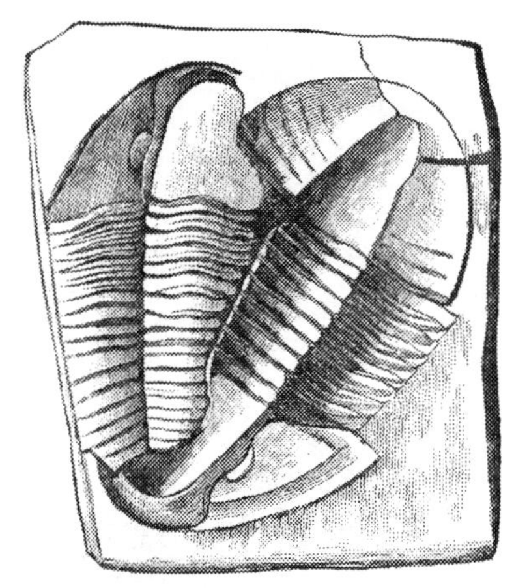

FIG. 2.—*Phillipsia æqualis,?* H. v. Meyer.
Culm, Aprath, Germany. Enlarged twice natural size. From the collection of John Edward Lee, Esq., F.S.A., F.G.S., of Torquay.

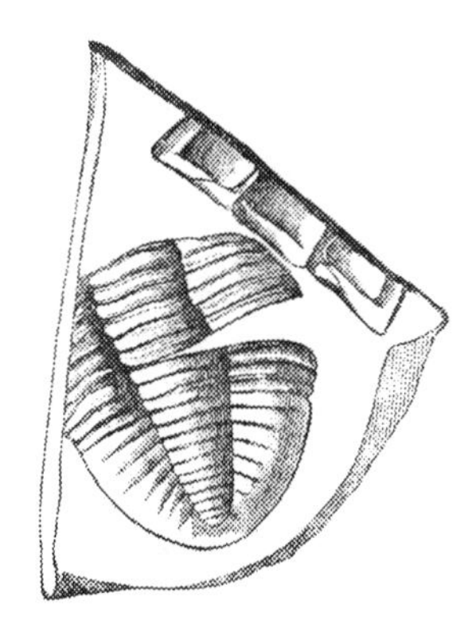

FIG. 3.—*Phillipsia*, sp.
From Culm-shale (Lower Carboniferous), near Marburg, Germany. Enlarged twice natural size. From the collection of John Edward Lee, Esq., F.S.A., F.G.S., Torquay.

We ought to add that von Meyer's figure of *? Calymene* (*Phillipsia*) *æqualis* ('Nova Acta,' vol. 15, 2 s., p. 100, Taf. 56, fig. 13) has no cheek-spines and no eyes, nor are any sutures shown in the head-shield.

The only other German Culm Trilobite we have had the opportunity to study is a compressed body and pygidium with the basal part only of the head-shield preserved in a piece of shale from near Marburg obtained by the Rev. G. F. Whidborne, M.A., F.G.S., who presented it to Mr. J. E. Lee, F.G.S., of Torquay, in whose cabinet I found it. (See woodcut fig. 3, p. 67.)

It agrees most nearly with our *Ph. Cliffordi* as regards the pygidium, which, however, consists of fewer segments, and the base of the head-shield shows that the Marburg specimen had not (apparently) any cheek-spines to the cephalon.

Formation.—Lower Culm.

Locality.—Waddon-Barton, near Chudleigh, Devon. From the Collection of Mr. J. E. Lee, F.S.A., F.G.S.

29. PHILLIPSIA MINOR, *H. Woodw.*, sp. nov. Pl. X, figs. 5, 6 *a*, *b*, 7, and 8 *a*.

This is the smallest Carboniferous Trilobite which I have studied, being only half the size of the smallest specimen of *P. Colei*, Pl. II, fig. 1.

Head-shield rounded in front, one-third broader than long; the glabella occupies one-third of its breadth, and is oval in outline, slightly broader in front, tumid, with distinctly marked basal lobes; lateral furrow indistinct; surface of glabella and free-cheeks covered with minute puncta; neck-lobe rather deep and prominent, free-cheeks having a furrow around the margin parallel to the border; angle of cheek produced into a short slightly curved spine.

Thorax consisting probably of nine segments; axis very distinct, forming one-third the entire breadth of body; axal furrows well defined, each pleura strongly grooved down the centre; extremities rounded. The eye is considerably larger in this than in the preceding species, but can only be distinctly seen on the free-cheek (fig. 7 *c*).

Pygidium one-fifth broader than long; the axis forms one-third of the breadth at the proximal border, but diminishes rapidly to a somewhat acute point at rather less than one-fourth of its entire length from the posterior margin. There are fourteen segments in the axis and ten lateral pleuræ which bifurcate as they approach the margin of the shield.

Hypostome (Pl. X, fig. 7 *h*). There seems but little reason to doubt that the detached free-cheek (*c*) and the hypostome (*h*) lying upon the same slab with the nearly entire specimen of *P. minor* really belong to one and the same specimen.

The hypostome is as broad as it is long, the anterior margin once attached to the underside of the front of the head-shield is rounded in contour, and expands laterally into two small lobes; posteriorly the hypostome is elongated into a pentangular lobe with a slightly raised margin, and a rounded central depression, the surface of which is striated.

The head of *P. minor* appears to us to be very distinct from any of the species of Carboniferous Trilobites hitherto figured, but the pygidium may be compared with *P. Eichwaldi* (Pl. IV, figs. 9 and 13); the axis, however, in *P. minor* tapers more rapidly to a point.

Formation.—Lower Culm.

Locality.—Waddon-Barton, Devonshire.

Figs. 6 and 7 are preserved in the British Museum (Natural History), and figs. 5 and 8 *a* are in the cabinet of Mr. J. E. Lee.

30. PHILLIPSIA CLIFFORDI, *H. Woodw.*, sp. nov. Pl. X, figs. 8 *b*, 9, 10, 11, 12.

The head of this species resembles that of *P. Leei*, but the cheek-spine is much shorter, and the lateral furrows on the glabella are more marked. The head is much distorted by cleavage, and its accurate description is attended with some difficulty, the head being imperfect. Thoracic segments unknown, probably nine in number.

Pygidium.—We are fortunately able to figure five pygidia of this species having well-marked characters of their own. The tail-shield is nearly twice as broad as it is long, the axis is one-third of its breadth at the proximal border, but rapidly diminishes to one-seventh at its bluntly-rounded extremity; here the shield is bordered by a wide margin covered with fine concentric striæ (being an impression of the underside); the border is one seventh the length of the pygidium, but diminishes in width laterally. The axis of the tail is composed of thirteen coalesced rings or somites, and has ten lateral pleuræ on each side; these bifurcate near their extremities as they approach the margin. There are no puncta, spines, or tubercles observable on this species.

The pygidium of *Ph. Cliffordi* agrees most in general facies with that of *Ph. Colei* (Plate II), but in the former the pleuræ are bifid at their extremities, not simple as in the latter species. It may also be compared with *Griff. Carringtonensis* (Pl. IX, figs. 6 *a*, *b*), but the Culm form is broader and shorter in proportion.

I have much pleasure in dedicating this species to Lord Clifford, of Ugbrook Park, Devonshire, upon whose estate the Culm is well exposed. These Trilobites were discovered by Mr. Lee on one of his lordship's farms at Waddon-Barton, and it was by his permission that Mr. Lee secured a quantity of the shale to break up, resulting in the discovery of the specimens here figured.

Formation.—Lower Culm-shale.

Locality.—Waddon-Barton, near Chudleigh, Devonshire.

Figs. 8 *b*, 9, 11, and 12, preserved in Mr. J. E. Lee's cabinet, Villa Syracusa, Torquay.

Fig. 10 preserved in the British Museum (Natural History).

31. Phillipsia articulosa, *H. Woodw.*, sp. nov. Pl. X, figs. 6 *c*, *d*, and 13.

This species is based upon three pygidia of Trilobites having a larger number of coalesced segments than any of the Culm specimens here noticed. We are unable to refer any cephalothorax as belonging to this form of pygidium, and can only therefore note its occurrence.

Pygidium one-fourth broader than long, axis one-third the entire breadth, consisting of seventeen coalesced segments which diminish rapidly in breadth to the extremity, which is bluntly rounded, and less than one-third the breadth the axis at the proximal end; axal furrows deeply marked. Pleuræ thirteen in number, terminating abruptly within the margin, which is finely striated; neither axis nor pleuræ have any ornamentation upon them.

This pygidium agrees most nearly with the preceding species, from which, however, it differs in possessing a greater number of coalesced somites, a character which seems sufficient to justify its separation.

Formation and locality the same as that of the preceding species.

These specimens are preserved in the British Museum (Natural History).

APPENDIX.

Note on the Nature of Certain Pores Observable in the Cephalon or Head-shield of some Trilobites. (Plates IV and IX.)

Among the numerous specimens of Carboniferous Trilobites which I have had the opportunity to examine during the last three years, many examples exhibit a peculiarity of structure which had already arrested the notice of such keen observers as Portlock, M'Coy, Oldham, Salter, Barrande, and Valerian von Möller. I allude to certain pores more or less well-marked, and placed usually one on either side of the glabella in the axal furrow, and upon the facial suture which separates the free-cheek from the fixed one, forming the margin of the glabella, and just in front of the compound eyes (see Plate IV, figs. 6, 8 and 10, p. 22).

In working at the Silurian Trilobites Professor F. M'Coy, in his 'Synopsis of the Silurian Fossils of Ireland, collected by Sir Richard Griffith' (Dublin, 1846, 4to, p. 43), writes, "I have observed in several Trilobites a peculiar pore situated in the furrow which separates the glabella from the cheeks near the anterior margin on each side, which seems to have escaped general notice, and which it is not impossible may be the remains of setaceous external antennæ." Professor M'Coy thinks these pores occupy just the position which the antennæ would have occupied, and that antennæ, being hollow organs, would leave a hole in the external integument if broken off. He then proceeds to observe that "in *Ampyx* these punctures are extremely remarkable and obvious, and it is only in this group that they have, to my knowledge, been observed by naturalists. Captain Portlock[1] having noticed them, but without any remark, in the description of his *A. Sarsii.* They are to be seen, but not exactly in the right place, in his figure of that fossil, in which they form two very deep, oblong punctures, communicating with the interior; they are situated in the furrow before-mentioned, about their own length within (or posterior to) the anterior margin. I have likewise observed them in all the species of the genus *Trinucleus.* In *T. Caractaci* they form two rather large circular punctures, as large as one of the punctures of the wings; they are in the

1 'Report on the Geology of the County of Londonderry, and of parts of Tyrone and Fermanagh' (Dublin, 1843), p. 261.

furrows before mentioned, close to the anterior margin of the cheeks, communicating with the interior. In *T. seticornis* they form two small circular punctæ within the margin; they are in the same furrows, penetrate to the interior, and are smaller than the punctures of the wings. In *T. radiatus* they hold the same position as in *T. Caractaci,* that is, very close to the anterior margin of the cheeks. They are rather larger in *T. elongatus,* in which they occupy the same position as in *T. Caractaci,* but are smaller, and placed rather more within the margin. In *T. fimbriatus,* though small, they are very conspicuous, being about the size of one of the punctures of the margin of the shield, and their own diameter within the margin. It is very remarkable that those organs are most obvious in what are considered *blind* Trilobites; whether it may be as a compensation for the want of eyes that they are furnished with better-developed antennæ, organs seeming so mysteriously to combine in themselves the exercise of all the senses, besides their own aeroscepsin, as Lehmann calls it, I am unable to conjecture."

He adds "that these punctures exist, although very much reduced, in Trilobites with eyes, I have also ascertained, as they are found, although exceedingly minute, in the common *Griffithides globiceps* of the Carboniferous Limestone, holding the same position as in the other genera alluded to, being seated in the same furrows, but farther back from the anterior margin."

Dr. T. Oldham, F.R.S., in the 'Journ. Geol. Soc.,' Dublin, 1846 (vol. iii, part 3, p. 189), writes in his description of *Griffithides globiceps:* "In the furrows which separate the cheeks and glabella, about half way between the front of the eye and the anterior margin, I have observed in all the tolerably preserved specimens which I have seen, a small hole or indentation. These are constant and therefore obviously connected with the structure of the creature, although I cannot offer an explanation of their use. They are similar to those noticed by Portlock in his *Ampyx Sarsii*" (op. cit., p. 261).

In 1847, Mr. J. W. Salter communicated a paper to the Geological Society, "On the Structure of *Trinucleus,* with remarks on the species" ('Quart. Journ. Geol. Soc.,' 1847, vol. iii, pp. 251—254), in which he observes: "The peculiar perforated border is the most interesting part of these animals, and I propose to examine it critically.

"The puncta are almost always arranged in radiating rows, three, four, or more holes in each row, and these being at equal distances they form concentric lines. In *T. granulatus,* two of these rows are separated by a furrow from the rest; in *T. seticornis,* three are distinct from the remaining two or three, by the front rows being sunk in a deep concentric furrow. Other modifications take place; in *T. fimbriatus,* the two front rows are turned downwards; lastly, in *T. ornatus*—for by that name we must call *T. Caractaci*—the dots occur most frequently in quincunx order, *i.e.* the radiant rows appear zigzag and not direct;

this appearance is due to the great obliquity of the ray. I wish to call attention to this, because I consider it enables us to understand the nature of the enigmatical puncta. If we suppose a head furnished with a produced membranous margin instead of a perforate one, we shall get at the explanation by supposing the membrane to collapse at regular intervals, become plicate, then perforate, and lastly, separate into linear processes. Now we have in *Harpes* the flat border, with rows of impressed puncta, which have not yet perforated the fringe. In *Trinucleus fimbriatus*, we have a plicate border, the thin interstices of which have contracted into pores, which is a step beyond the simple perforation in linear series exhibited by *T. ornatus*. Lastly, in *Ceraurus* (*Acidaspis*, Murch.), we have the structure completed, the linear processes being quite separated into spines. This structure is not anomalous, for the cellular membrane which forms the inner peristome of the mass passes through an exactly similar course, becoming in different species perforate, in others separated into distinct teeth.

"That these perforate or spinous fringes are not essential, but only supplementary parts of the head, may easily be shown by the fact that the width of the head, without the fringe, is exactly that of the body, and when the animal is doubled up the fringe projects freely on all sides. We still require to find anomalous specimens in which all or some of the above modifications, plications, perforations, or partially cleft borders, may be exhibited together, in order to demonstrate the supposed origin of the structure" (p. 253).

In another paragraph in the same paper Salter refers to the discovery by himself and Emmrich of the facial suture in *Trinucleus ornatus*. "Its course," he states, "is obliquely upwards from the eye tubercle to the upper end of the glabella, where it appears to terminate in a solitary deep perforation, similar to those which surround the head" (*op. cit.*, p. 251). It is to these pair of deep puncta, one on either side of the head, that we would specially draw attention.

Later on (in 1852) M. Barrande, in his "Système Silurien de la Bohême" (vol. i, *Trilobites*, p. 230), thus wrote: "On the antennæ of Trilobites:—Prof. M'Coy, in his work already cited (p. 43), announces what he deems to be the discovery of the remains of antennæ in the form of a deep pore on either side of the frontal lobe in the groove which surrounds the glabella. He enumerates the genera *Trinucleus*, *Acidaspis*, *Calymene*, *Ampyx*, *Griffithides*, &c., in all of which this observer states he has succeeded in tracing these organs. We have also remarked these puncta long since in a variety of Trilobites from Bohemia, notably in *Calymene*, *Trinucleus*, *Cheirurus*, but we have been led to offer a different interpretation to that indicated above.

"First, we have to observe that if this cavity does not appear to be anything else but a pore in those species with a border, as *Trinucleus*, it assumes larger and larger dimensions as it approaches the edge of these Trilobites, and forms a funnel-shaped opening nearly 2 mm. in diameter in the head of *Cheirurus claviger*."[1]

"When the shell exists, as we have seen it in specimens of *Calymene Baylei*, *Cheirurus gibbus*, &c., it is bent inwards, as a funnel-shaped depression. We have thought that this bending inwards of the shell was simply designed to afford points of attachment for the muscles of the jaws, and that they had the same origin as the similar indentations which we have indicated in the pleuræ of various species of Trilobites. We have observed the conformity of the position of these pits in the depression of the axial furrows of the thorax as in the head. We recall again the indentations of the shell on the glabellal furrows and on those of the axis of the pygidium in *Dalmannia.*

"We were satisfied with these analogies, and we sought no other explanation for these little indentations. But after having read the opinion of Prof. M'Coy, we have studied all the specimens afresh which might serve to elucidate this question, and we have found a portion of a *Cheirurus gibbus,* which seems to explain the matter satisfactorily to our views. This fragment is broken along the line of the dorsal groove and the length of the glabella, exposing to view one of the alæ of the hypostome *in sitû.* This wing of the hypostome fits at its extremity to the interior of the little funnel-shaped projection formed upon the underside of the head-shield. These details are very distinct, owing to the very perfect preservation of this specimen with its shell. After this we cannot accept any other interpretation for the genus *Cheirurus* than that which we have given. We leave to savans the task of pointing out the value of these analogies, and they may be able to apply this method of explanation to other genera mentioned by Prof. M'Coy" (*op. cit.*, p. 230).

Valerian von Möller, in his paper ("Ueber die Trilobiten der Steinkohlenformation des Urals") contributed to the 'Bulletin de la Société Impériale des Naturalists de Moscou," 1867, p. 44, notices these same pores in the head of *Phillipsia Eichwaldi* as "very distinct deep funnel-shaped openings, which run a little obliquely and enter into the underside of the cephalothorax." He cites the opinion of M'Coy and of Barrande, and says, in conclusion, "I quite agree with the observations of Barrande, and I feel sure the more one examines these indentations the more one feels satisfied that they are only superficial openings."

I have carefully examined these puncta in the Carboniferous Trilobites, and have sought for some satisfactory explanation in various other members of the class Crustacea, both recent and fossil.

[1] This is not shown in Barrande's figures of *Cheirurus claviger*, pl. 40, figs. 1—12, nor in *Ch. gibbas*, figs. 35—39, but is seen in pl. 41, fig. 17; it is also well shown in *Trinucleus ornata*, pl. 29, figs. 1—8, and in *Placoparia Zippei*, pl. 29, figs. 30—34, and in *Calymene Baylei*, pl. 43, fig. 49, and in many others not referred to by Barrande.

And first I would observe, in reference to the varied forms of *Trinucleus*, with corrugated, punctate, perforate, and serrated margins to the head-shield, that we may see in the larval "king crab" or "horse-shoe crab" (*Limulus polyphemus*) of North America just before hatching, that the head-shield retains indications, in the well-marked fimbriated hepatic lobes, of the presence of a once divided border, corresponding with the five or six distinct cephalic somites. This fimbriated margin to the cephalic shield, seen in the young of *Limulus*, is observable also in the head-shield of *Hemiaspis limuloides* from the Lower Ludlow of Shropshire. As to the explanation given by M. Barrande for the series of depressions, or invaginations, of the crust, of various Trilobites which are seen to correspond with each thoracic somite, and are placed exactly in the depression of the furrow on each side of the dorsal axis, and are even observable in the axal furrows of the pygidium of *Dalmannia*, we agree with him entirely, and there can be, I think, no doubt whatever that they are perfectly homologous with the similar pits or indentations of the crust, observable in the thoracico-abdominal shield of *Limulus*, which gives rise to six powerful calcareous processes within (*Apodemata*, Milne Edwards; *Entapophyses*, Owen) for the attachment of the muscles required for the locomotory organs.

But whether the two pits or pores placed one on each side of the glabella, in front of the compound eyes, are of the same nature (as Barrande assumes them to be) is an open question. That they are not the remains of *Antennæ*, as supposed by Prof. M'Coy, I think we may feel quite satisfied, as we know of no Crustacean whatever with antennæ arising from the *dorsal* aspect of the cephalic shield in the manner supposed by M'Coy.

The antennæ are either on the *margin* of the head as in many Isopods (see Plate IX, figs. 8 and 9), or *beneath* as in *Limulus*, *Apus*, and other forms of Entomostraca.

Mr. John Young, F.G.S., of the Hunterian Museum, in the University of Glasgow, a most careful observer, and to whom I am indebted for the loan of many interesting specimens of Carboniferous Trilobites, several of which illustrate this point in an admirable manner, suggests, in a note to me, that he has long thought that these puncta might possibly prove to be *ocelli*. Whether this be the correct interpretation or not, at any rate it seems a more plausible hypothesis than that suggested by Prof. M'Coy.

It is true that such an association of larval eye-spots with compound eyes has not as yet been met with in any of the TRILOBITA, but it occurs in the MEROSTOMATA, both recent and fossil (*Pterygotus*, *Slimonia*, and *Limulus*), and its discovery would be hailed as further evidence in favour of the undoubted close affinities of both these groups with the *Scorpionidæ*, and so with the ARACHNIDA.

We have noticed two instances of pores in the *Isopoda*, and it is probable that the researches into the "Challenger" collections now in progress will reveal many others in this interesting group. In one instance, *Sphæroma* (Pl. IX, fig. 9), the pore seems really to be on the line of the suture of the glabella, the free cheeks of the Trilobite being represented in the Isopod by the first thoracic somite which wraps around the sides of the head in a similar manner.

In *Serolis* (Pl. IX, fig. 8) the pore, which forms rather an elongated slit (*p*), is in the centre of the margin of the first thoracic somite, at a distance from the compound eye, and has no relation to any suture. These puncta may be, like the *fenestræ* in the head of *Blatta orientalis*, either rudimentary ocelli or the seat of some other nerve-sense, and may have been, as in *Blatta* and in *Serolis*, covered with a thin transparent portion of the integument, which served either as a simple eye, a tympanum, or an olfactory pore. We have referred to these *fenestræ* in the head of *Blatta* because they are placed, like those in the Trilobites, *on a suture* of the head, and in front of the compound eyes.

We hope to have some further information to offer on this subject upon a future occasion, meantime other workers may be able to assist with suggestions and specimens to elucidate this interesting point, for which we shall be only too thankful.

On the OCCURRENCE *of the Genus* DALMANITES *in the* LOWER CARBONIFEROUS ROCKS *of* OHIO. By Prof. E. W. CLAYPOLE, B.A., B.Sc. (Lond.), F.G.S., of Buchtel College, Akron, O., United States. (See 'Geol. Mag.,' 1884, p. 303.)

Of the abounding Trilobites which mark the faunas of the Lower and Middle Palæozoic rocks few survive into the Upper Palæozoic. Three genera, if indeed they really deserve that name, have been described from the Carboniferous beds in England—*Phillipsia*, *Griffithides*, and *Brachymetopus*. Only the first of these is yet known to occur in American Palæozoic strata. But on the other hand two species of *Proetus* have been announced from America—a genus not yet recognised in England.[1] It is true that the distinctions are so slight that possibly these last might be as correctly referred to *Phillipsia* as to *Proetus*. As they stand, however, the distribution of the North American Carboniferous Trilobites is as follows:

Distribution of North American Carboniferous Trilobites.

COAL-MEASURES	*Phillipsia*	*Cliftonensis*, Shum. 1858.
	,,	*major*, Shum. 1858.
	,,	*Sangamonensis*, M. and W. 1865.
	,,	*scitula*, M. and W. 1865.
CHESTER GROUP	,,	*Meramecensis*, Shum. 1855.
	,,	*Stevensoni*, Meek, 1871.
KEOKUK GROUP }	,,	*Lodiensis*, Meek, 1875.
}	*Proetus*	[2] *auriculatus*, ? Hall, 1861.
CUYAHOGA SHALE }	*Phillipsia*	*bufo*, M. and W. 1870.
	,,	*Portlocki*, M. and W. 1865.
BURLINGTON GROUP	,,	*insignis*, Win. 1863.
	,,	*tuberculata*, M. and W. 1870.
KINDERHOOK	,,	[3] *Doris*, Win. 1865.
	,,	*Rockfordensis*. Win. 1865.
	,,	[4] *Missouriensis*, Shum. 1858.
	,,	*Tennesseensis*, Win. 1869.
	Proetus	*ellipticus*, M. and W. 1865.

[1] See *ante*, p. 57. Prof. Claypole had not seen my notice of *Proetus*, from the Carboniferous Limestone of Ireland, when this was written.—H. W.

[2] Assigned in Miller's 'Catalogue of N. A. Palæozoic Fossils' to the Chemung Gr.

[3] Omitted from Miller's 'Catalogue.'

[4] Assigned to the Coal-Measures in Miller's 'Catalogue.'

Two other Canadian species of uncertain horizon complete the list:

Phillipsia Howi, Billings, 1863.
„ *Vindobonensis*, Hartt, 1868.

On both sides of the Atlantic, therefore, so far as I am aware, the Carboniferous Trilobites without exception belong to the type possessing a pygidium with definite, even outline, like that of the two genera given above. It is true, Meek says, in his description of *Phillipsia Lodiensis* ('Pal. of Ohio,' vol. ii, p. 324), "The fimbriated character of the posterior and lateral margins of the pygidium is very peculiar and hitherto unknown, I believe, in either of the above-mentioned genera" (*Phillipsia* and *Griffithides*), "though it occurs in one section (*Phaethon*) of the allied genus *Proetus;* hence, it is possible that our species should be called *Proetus* (*Phaethon*) *Lodiensis,* as it would not be very surprising that this genus should be found in this oldest member of the Carboniferous, though hitherto, I believe, only known in the Silurian and Devonian."[1]

The crenate character of the margin of the pygidium here alluded to must have been exceedingly slight in the typical specimen, as not a trace of it appears in the figure (pl. xviii, fig. 3). Prof. Meek says of this feature, "The segments are continued down upon and across the sloping border, at the edge of which they terminate in little pointed projections so as to present a fimbriated appearance around the posterior and lateral margins. (This latter character is not represented in the figure.)"

It is obvious from the terms here employed that the crenation alluded to in no wise resembles the pointed and almost spinous margin of the species described below. I may add that specimens of *Phillipsia Lodiensis* from the Cuyahoga shale of this county (Summit), only a few miles from the locality of Prof. Meek's type, show no perceptible crenation.

In regard to the two species of *Proetus* on the list given above, *P. ellipticus* and *P. auriculatus,* a few words may be added. I have not seen specimens of either, but the description of the former by its authors shows that it differs very slightly, almost imperceptibly, from *Phillipsia.* They remark in conclusion, "It is very probable that we should call this species *Phillipsia elliptica,* as it seems to present most of the characters of that genus. Unfortunately, the characters distinguishing these groups seem not to have been very clearly pointed out." ('Proc. Acad. Nat. Sci. Phil.', 1865, p. 267-8.)

The pygidium of *Proetus auriculatus,* Hall, is not certainly known. That which is supposed to belong to this species is "marked by seven or eight ribs

[2] In writing this sentence, 1875, Prof. Meek seems to have forgotten his own ***Proetus ellipticus*** of 1865, and Prof. Hall's ***Proetus auriculatus*** of 1861, the former from the Kinderhook, and the latter from the Waverley Sandstone.

terminating in a wide-spreading border." ('15th Report on State Cabinet of New York,' 1862, p. 107.)

In *Proetus Verneuili*, as figured by Prof. Hall in his Illustrations of Devonian Fossils (pl. xv, fig. 18), the lateral ribs extend partly across the border, forming a line of marginal tubercles, but there is no crenation of the edge. This is the nearest approach to the appearance described by Prof. Meek (loc. cit.) with which I am acquainted.

It follows, therefore, that with the single possible exception of *Phillipsia Lodiensis*, as described by Prof. Meek, no Trilobite has been announced from the American Carboniferous rocks in which the pleural ridges of the pygidium extend beyond the border. The occurrence, therefore, of the form (*Dalmanites? Cuyahogæ*) figured and described below (fig. 6) is of considerable interest to palæontologists. It was found in the Cuyahoga Shale, the uppermost member of the Lower Carboniferous system in Northern Ohio.

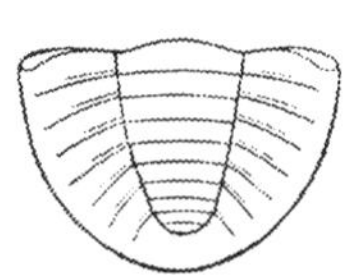

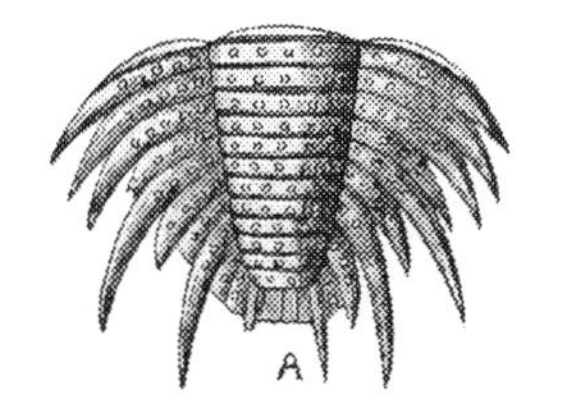

FIG. 4. *Griffithides globiceps*, De Kon. (twice nat. size). 5. *Phillipsia Derbiensis*, Martin (twice nat. size). 6. *Dalmanites? Cuyahogæ*, Claypole (twice nat. size). 7. *Phillipsia Lodiensis*, Meek (twice nat. size). 8. *Brachymetopus discors*, M'Coy. (five times nat. size).

Pygidia of species of Carboniferous Limestone Trilobites, *Griff. globiceps*, *Ph. Derbyensis*, *Ph. Lodiensis*, and *Brach. discors.*, compared with the pygidium of *Dalmanites Cuyahogæ*.

DALMANITES? CUYAHOGÆ, Claypole. Geol. Mag., Decade iii, vol. i, p. 306, 1884. (Woodcut, Fig. 6.)

Head and thorax unknown. Pygidium about as broad as long, exclusive of the spinous processes to be mentioned below; distinctly trilobate. Middle lobe occupying a full third of the breadth, extending nearly to the hind extremity, distinctly separated by a furrow from the lateral lobes, and containing eleven or twelve ridges or partial rings. Lateral lobes as wide as the medial lobe, their segments or ribs produced for the most part about half their length beyond the marginal line formed by their union and ending in points, the third, seventh, and ninth produced to double the distance and having the appearance of spines. The third pair curve backward so that their points are level with the hind end of the pygidium. No marginal tract. Whole surface set with tubercles, of which there are five on each of the three parts of the first ring, the number gradually diminishing toward the hind extremity; part of segment prolonged beyond the marginal line free from tubercles.

Horizon and Locality.—Cuyahoga Shale of Lower Carboniferous, Akron, Ohio.

It will of course be understood that the reference of the specimen to the genus *Dalmanites* is provisional only, and must await for confirmation the discovery of other parts of the carapace. I may add that since the above description was written I have found a second specimen and seen a third, both very imperfect, and only sufficient to support the statement of the existence of this type in the Lower Carboniferous rocks.

On Tomocaris Piercei, a recent Isopod, *dredged* East of Cape Frio, *by the late* Prof. L. Agassiz, *offering affinities with the* Trilobita.

The following letter, from the late Prof. Agassiz to Prof. Pierce, of Harvard College,[1] seems deserving of notice as recording the discovery of an abnormal form of existing Isopod, having many points of affinity with the extinct Trilobita.[2]

"In my first letter to you concerning deep-sea dredging, you may have noticed the paragraph concerning Crustacea in which it is stated that among these animals we may expect 'genera reminding us of some Amphipods and Isopods aping still more closely the Trilobites than *Serolis*.' A specimen answering fully to this statement has actually been dredged in forty-five fathoms, about forty miles east of Cape Frio. It is a most curious animal. At first sight it looks like an ordinary Isopod, with a broad, short, flat body. Tested by the characters assigned to the leading groups of Crustacea, whether we follow Milne Edwards' or Dana's classification, it can, however, be referred to no one of their orders or families. As I have not the works of the authors before me, I shall have to verify more carefully these statements hereafter; but I believe I can trust my first inspection. The general appearance of my new Crustacean is very like that of *Serolis*, with this marked difference, however, that the thoracic rings are much more numerous and the abdomen or pygidium is much smaller. It cannot be referred to the Podophthalmians of Milne Edwards (which correspond to the Decapods of Dana) because it has neither the structure of the mouth, nor the gills, nor the legs, nor the pedunculated eyes of this highest type of the Crustacea, nor can it be referred to the Tetradecopods of Dana (which embrace Milne Edwards' Amphipods and Isopods), because it has more than seven pairs of thoracic limbs; it cannot be referred to the Entomostraca, because the thoracic are all provided with locomotive appendages of the same kind. But it has a very striking resemblance to the Trilobites; it is, in fact, like the latter, one of those types combining the characteristic structural features of other independent groups which I have first distinguished under the name of synthetic types. Its resemblance to the Trilobites is unmistakable, and very striking. In the first place the head stands out distinct from the thoracic regions, as the buckler of

[1] 'Canadian Naturalist,' vol. vi, new series, 1872, p. 359.

[2] I am indebted to Mr. F. Beddard, Prosector of the Zoological Society of London, for calling my attention to *Tomocaris*. Mr. Beddard informs me he has written to Cambridge, U.S., to inquire about the specimen, but it has unfortunately been mislaid or lost, and no other has since been obtained.

Trilobites; and the large, kidney-shaped, faceted eyes recall those of Calymene; moreover, there is a facial suture across the cheeks, as in Trilobites, so that, were it not for the presence of the antennæ, which project from the lower side of the anterior margin of the buckler in two unequal pairs, these resemblances would amount to an absolute identity of structure. As it is, the presence of an hypostome in the same position as that piece of the mouth is found in Trilobites, renders the similarity of this extinct type of Crustacea still more striking, while the antennæ exhibit an unmistakable resemblance to the Isopods.

"In view of the synthetic character of these structural features it should not be overlooked that the buckler of our new Crustacean, for which I propose the name *Tomocaris Peircei*, extends sideways into a tapering point, curved backwards over the first thoracic ring, as is the case with a great many Trilobites. The thorax consists of nine rings, seven of which have prominent lateral points, curved backwards like the pleuræ of *Olenus*, *Lichas*, &c. The sixth ring is almost concealed between the fifth and seventh, and is destitute of lateral projections, as is also the ninth. Those rings are distinctly divided into three nearly equal lobes by a fold or bend on each side of the middle region, so that the thorax has the characteristic appearance of that of the Trilobites, to which the latter owes its name. The legs are very slender, and resemble more those of the Copepods and Ostracods than those of any other Crustacea. There are nine pairs of them, all alike in structure, six of which, however, the anterior ones, are larger than the three last, which are also more approximated to each other. Besides the legs there is a pair of maxillipeds attached to that part of the buckler which extends back to the facial suture. These maxillipeds resemble the claw of a Cyclops. All these appendages are inserted in that part of the rings corresponding to the bend of the thoracic lobes; so that if there exists a real affinity between the Trilobites and our little Crustacean—and their resemblance is not simply a case of analogy—we ought hereafter to look to a corresponding position for the insertion of the limbs of Trilobites. I do not remember with sufficient precision what Billings, Dana, and Verrill have lately published concerning the limbs of Trilobites to say now what bearing the facts described above may have upon the subject as lately discussed in 'The Journal of Science.' But of one thing I am satisfied, since I have examined the *Tomocaris Peircei* that Trilobites are not any more closely related to the Phyllopods than to any other Entomostracæ, or to the Isopoda. In reality the Trilobites are like *Tomocaris*, a synthetic type, in which structural features of the Tetradecapods are combined with characters of Entomostraca and other peculiarities essentially their own.

"The pygidium or abdomen of *Tomocaris* is very like the abdomen of the ordinary Isopods with an articulated oar attached sideways, and leaf-like respiratory organs upon the under side. The whole pygidium is embraced between the last curved points of the side of the thorax. Owing to these various combinations I would expect in Trilobites phyllopod-like respiratory appendages under the pygidium only, and slender, articulated legs, with lateral bristles under the thorax, so thin and articulated by so narrow a joint as easily to break off without leaving more than a puncture as an indication of their former presence. It is impossible to study carefully the synthetic types without casting a side glance at those natural groups, which, without being strictly synthetic themselves, have nevertheless characters capable of throwing light upon the whole subject. And in this connection I would say a few words of *Apus* and *Limulus*. If I remember rightly, Milne Edwards considers the shield of *Limulus* as a cephalo-thorax in which the function of chewing is devolved upon the legs, while he regards the middle region as an abdomen, and the sword-like tail as an appendage *sui generis*. In the light of what proceeds I am rather inclined to consider the cephalic shield of *Limulus* as a buckler homologous to that of the Trilobites, and the middle region as a thorax in which the rings show unquestionably signs of a division into lobes as in Trilobites. The tail would then answer to the pygidium. *Apus* should be compared with the other Crustacea, upon the same assumption as *Limulus*.

L. AGASSIZ."

" *Rio, on board the 'Hassler,'*
" *Feb.* 12*th*, 1872.

LIST OF PLATES.

LIST OF WOODCUTS.

INDEX.

PLATE I.

Carboniferous Trilobites.

Phillipsia Derbiensis, Martin, sp. 1809. (P. 12.)

Figs. 1 *a* and 1 *b*.—Head, from the Carboniferous Limestone, Bolland, Yorkshire. (Gilbertson Coll.) Magnified four times. 1 *a*, top view; 1 *b*, side view of same. Original specimen in the British Museum (Nat. Hist.), Cromwell Road, S.W.

Figs. 2 *a* and 2 *b*.—A complete specimen from the Carboniferous Limestone, Longnor, Staffordshire. Magnified three times. 2 *a*, side view ; 2 *b*, top view of same. Original specimen in the Museum of Practical Geology, Jermyn Street, S.W.

Fig. 3.—Detached head, from the Carboniferous Limestone, Settle, Yorkshire. Magnified three times. Original specimen in the Woodwardian Museum, Cambridge.

Figs. 4 *a*.—Entire specimen; and 4 *b*.—Hypostome of same, from the "Rotten stone" band of the Carboniferous Limestone, Matlock, Derbyshire. Fig 4 *a*, magnified twice, and 4 *b* (the hypostome of 4 *a*), magnified four times. Original specimen in the British Museum (Nat. Hist.).

Fig. 5.—Detached head, from the Carboniferous Limestone of Castleton, Derbyshire. Original specimen in the collection of the Rev. E. Oldridge de la Hey, M.A., Marple, Cheshire. Magnified three times.

Fig. 6.—*Phillipsia Derbiensis*, Martin, restored outline.

Fig. 7.—*g*, Outline of glabella; *c*, outline of free cheek.

Fig. 8 *a*, *b*.—Top view and section in outline of one of the thoracic somites.

Fig. 9.—Outline restoration of pygidium, enlarged.

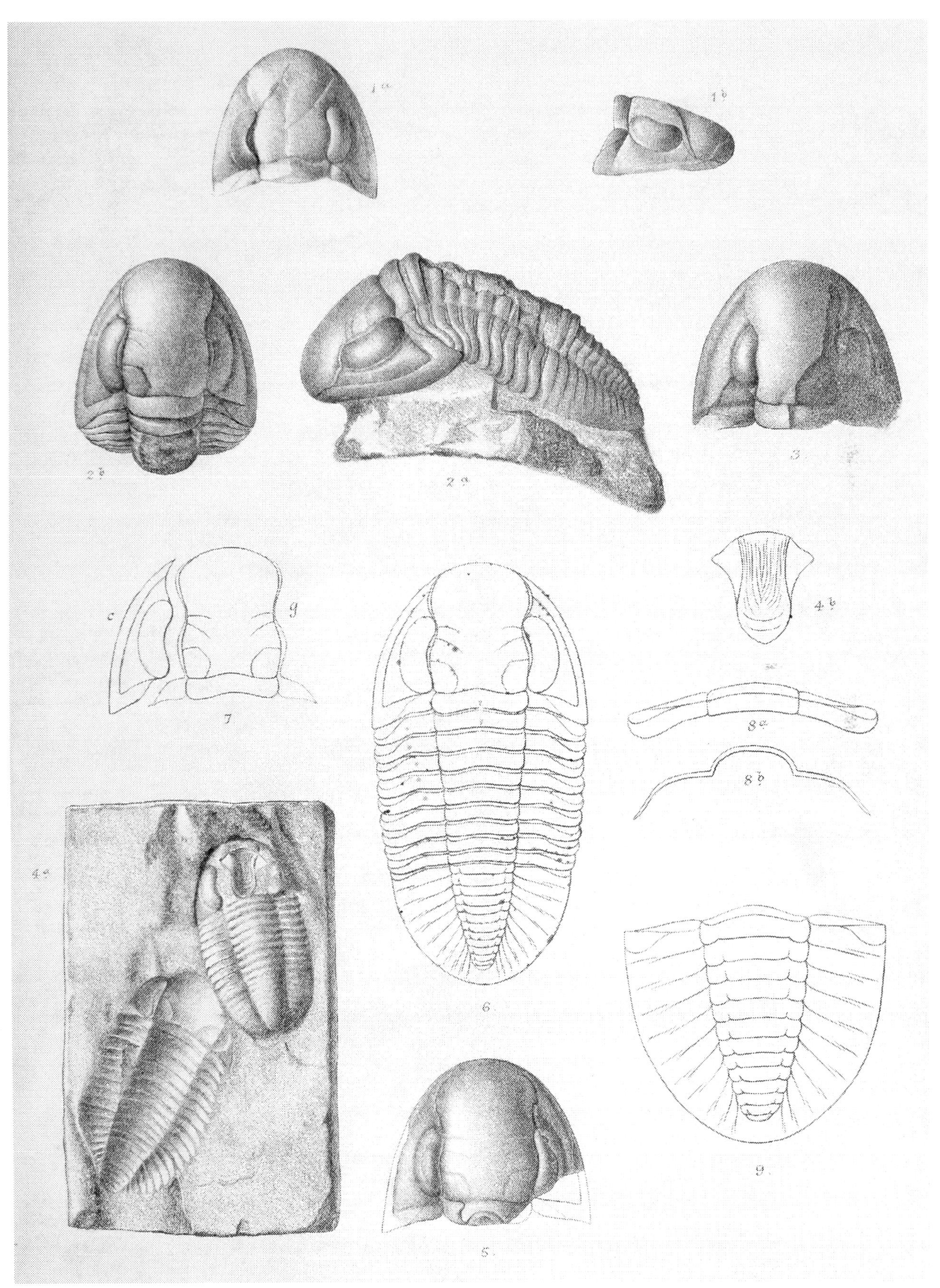

G. M. Woodward del. et lith.

West Newman & Co imp.

PLATE II.

CARBONIFEROUS TRILOBITES.

Phillipsia Colei, M'Coy. 1844. (P. 16.)

FIG. 1.—A small and rather mutilated specimen, wanting the cheek or genal portion of the head and some of the thoracic somites. The eye on the left side is well preserved. Magnified six times natural size.

FIG. 2.—A larger and more perfect specimen; the genal portion of the head is not preserved clearly, and the eyes are wanting. Magnified twice natural size.

FIG. 3.—Another example, wanting the lateral portion of the head and the eyes; the thoracic segments on one side are very well preserved. Magnified three times natural size.

FIG. 4.—This specimen is preserved as an intaglio only, but shows the glabella, the thoracic somites, and the pygidium united. Magnified three times natural size.

FIG. 5.—A very complete caudal shield or pygidium. Magnified twice natural size.

FIG. 6.—A detached hypostome. Magnified four times.

All the above specimens are from the Museum of the Geological Survey of Ireland, Dublin, and were obtained from the black Carboniferous Shale, occurring in rocks in a stream opposite Flax Mill, north-east of Ballintra and Carrickbreeny, Donegal, Ireland.

FIG. 7.—*Phillipsia Colei*, M'Coy. Restored outline, three times enlarged.

FIG. 8.—Outline of glabella, *g*. Cheek, *c*, separated to show the suture.

FIG. 9 *a*, *b*.—One of the thoracic ribs drawn in outline, seen from above, and in profile.

FIG. 10.—The pygidium, drawn in outline.

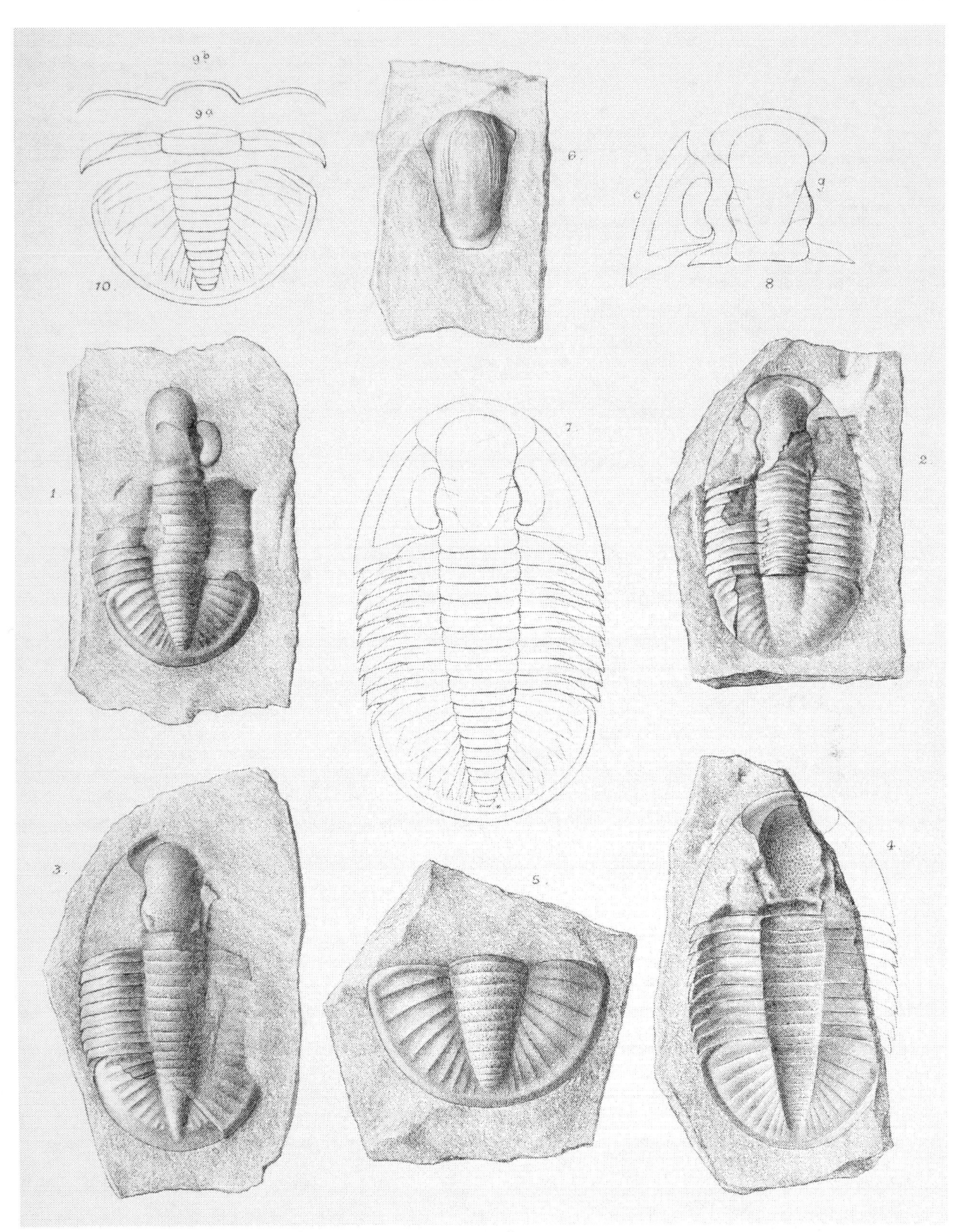

G. M. Woodward del. et lith.

West Newman & Co imp.

PLATE III.

Carboniferous Trilobites.

Figs. 1—8.—*Phillipsia gemmulifera,* Phillips, sp., 1836. (P. 17.)

Fig. 1.—A very perfect and entire specimen in hard crystalline Carboniferous Limestone, Kildare, Ireland. Magnified three times natural size. Original specimen in the Museum of Practical Geology, Jermyn Street.

Fig. 2.—A perfect detached head (one of the eyes showing the facetted surface very clearly), preserved in dark crystalline Carboniferous Limestone, St. Doolagh, Dublin. Magnified twice natural size. Original specimen in the Museum of the Geological Survey of Ireland, Dublin.

Fig. 3.—An entire thorax and abdomen, but without a head, from the Carboniferous Limestone, Clitheroe, Lancashire. Magnified twice natural size. Original specimen in the collection of John Aitken, Esq., of Sandfield, Urmston, Manchester.

Fig. 4.—A very perfect detached pygidium, preserved in white crystalline Carboniferous Limestone; said to be from Derbyshire (more probably from Settle, Lancashire). Magnified twice natural size. Original specimen in the Woodwardian Museum, Cambridge.

Fig. 5.—Another well-preserved detached pygidium, in crystalline Carboniferous Limestone from Bolland, Yorkshire. Magnified twice natural size. Original specimen in the British Museum (Nat. Hist.), Cromwell Road.

Fig. 6.—*Phillipsia gemmulifera,* Phil., sp. Restored outline.

Fig. 7 *a* and *b*.—One of the thoracic somites, drawn in outline.

Fig. 8.—One of the eyes enlarged six times to show the facets. From a specimen in Mr. John Rofe's collection, now in the British Museum (Nat. Hist.), Cromwell Road.

Figs. 9—14.—*Phillipsia truncatula,* Phillips, sp., 1836. (P. 21.)

Fig. 9.—A detached head, from black Carboniferous Limestone, with *Fenestella* from Hook Point, Co. Wexford, Ireland. Drawn of natural size. Original specimen in the Museum of the Geological Survey of Ireland, Dublin.

Fig. 10.—A second detached head, from the same formation and locality. Natural size.

Figs. 11 and 12.—Two detached pygidia, from the same locality. Original specimens of figs. 10, 11, and 12, in the Woodwardian Museum, Cambridge.

Fig. 13.—*Phillipsia truncatula,* Phil., sp. Restored outline of head.

Fig. 14.— " " Restored outline of pygidium.

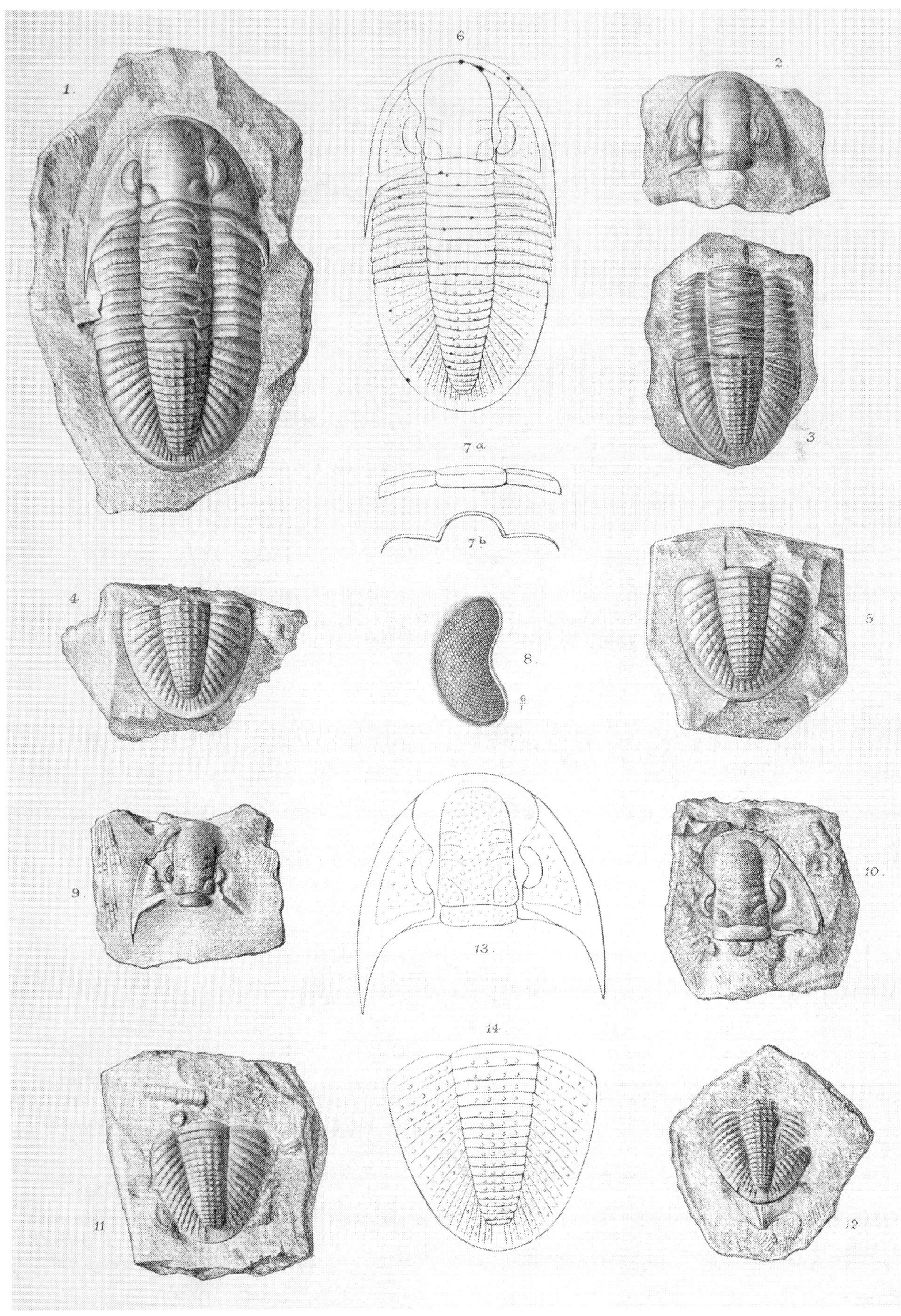

G. M. Woodward del. et lith.

West Newman & C° imp.

PLATE IV.

Carboniferous Trilobites.

Fig. 1.—*Phillipsia Eichwaldi*, var. *mucronata*, M'Coy. An almost complete specimen showing the mucronate tail, and the long cheek-spines so characteristic of this species. The right side is however injured. From the black Carboniferous Limestone of Wilkiestone, Fife. Drawn natural size. Original specimen in the cabinet of Dr. R. H. Traquair, F.R.S., of Edinburgh. (P. 23.)

Fig. 2.—*Phillipsia Eichwaldi*, Fischer, sp. A nearly perfect specimen, agreeing closely with Fig. 1 in all particulars of structure save that the mucro of the pygidium is wanting. From the Lower Limestone series, Newfield Quarry, near High Blantyre, Lanarkshire. Magnified twice the natural size. Original specimen in Mr. J. Young's collection, Glasgow. (P. 22.)

Fig. 3.—*Ph. Eichwaldi*, var. *mucronata*, M'Coy. A detached head magnified twice the natural size, from the same horizon and locality as Fig. 1. From Dr. Traquair's cabinet. (P. 23.)

Fig. 4.—*Ph. Eichwaldi*, Fischer, sp. A detached hypostome, believed to belong to this species. Enlarged three times the natural size. From the Upper Limestone series, Dalry, Ayrshire. Original specimen in Mr. John Young's collection. (P. 22.)

Fig. 5*a*.—A nearly entire rolled-up specimen, the cheek-spines and extremity of the pygidium are wanting; the facets of one eye are beautifully preserved. (P. 22.)

Fig 5*b*.—Thoracico-abdominal segments of same fossil. The original specimen in Mr. J. Thomson's collection, Glasgow.

Fig. 6.—Detached glabella, showing pores (*p*) on each side of head in front of the eyes. From same locality and collection as Fig. 4. Enlarged twice natural size. (P. 22.)

Fig. 7.—Detached hypostome. From the Upper (Carboniferous) Limestone Series, Dalry, Ayrshire. Magnified four times natural size. Original specimen in the collection of Mr. R. Craig, Langside, Beith, Ayrshire. (P. 22.)

Fig. 8.—Underside of head, exhibiting the pores (*p*) upon the underside of the margin of the glabella, at its union with the cheeks on each side, in front of the compound eyes. Lower (Carboniferous) Limestone Series, Sculliongour near Lennoxtown, Campsie, Sterlingshire. Magnified twice natural size. Original specimen in Mr. John Young's collection, Glasgow. (P. 22.)

Fig. 9.—Part of thorax and perfect rounded pygidium. From the same locality and formation as Fig. 7. Magnified twice natural size. Original specimen in Mr. J. Young's collection. P. 22.

Fig. 10.—Head, closely associated with a rounded pygidium, like Figs. 9 and 13. Showing pores (*p*.) at sides of glabella as in Figs. 6 and 8. Enlarged twice natural size. From the same locality and formation as Fig. 9. Original specimen in Mr. Young's collection. (P. 22.)

Fig. 11.—Free cheek showing the compound eye and the long cheek-spine, magnified twice natural size. From Shale below Fourth Limestone, Dalry. Original specimen in Mr. J. Smith's collection, Stobs, Kilwinning. (P. 22.)

Fig. 12.—*Ph. Eichwaldi*, var. *mucronata*, M'Coy. Mucronated pygidium (twice natural size) from Shale below Upper Limestone, Garple Water, Muirkirk. Mr. J. Smith's collection. (P. 23.)

Fig. 13.—*Ph. Eichwaldi*, Fischer, sp. Rounded pygidium like Fig. 9, and from the same locality and collection as Figs. 7 and 9. Magnified twice natural size. (P. 22.)

Fig. 14*a*, *b*, and *c*. A nearly entire and rolled-up specimen, the cheek-spines are broken off, but were evidently long as in Figs. 1 and 2. The tail is non-mucronate, as in Fig. 2. Both specimens are from the same locality and collection. *a*, View of head seen from above; *b*, profile of same; *c*, the pygidium of same specimen. Magnified twice natural size. (P. 22.)

Fig. 15.—*Ph. Eichwaldi*, var. *mucronata*, M'Coy. Outline restoration; (*p*) position of pores on glabella. (P. 23.)

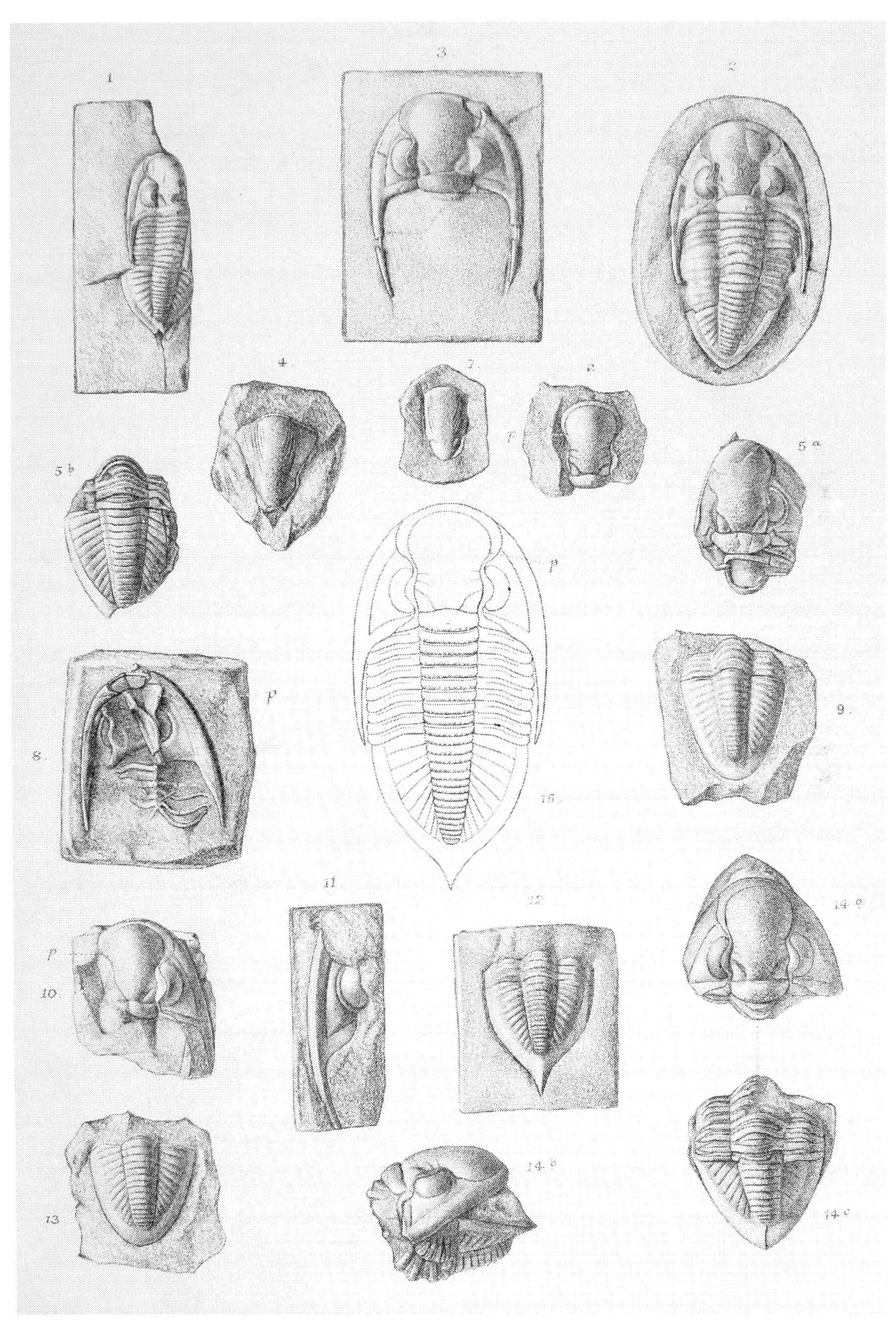

G. M. Woodward del. et lith.

West Newman & Co. imp.

PLATE V.

Carboniferous Trilobites.

Griffithides (Phillipsia) seminiferus, Phillips, sp. 1836. P. 28.

Fig. 1.—Intaglio of a nearly perfect specimen associated with a detached pygidium and fragment of another specimen preserved in the same piece of matrix. Drawn of natural size.

Fig. 2—Copy of a cast in relievo. Taken from Fig 1.

Fig. 3.—Another specimen, also preserved in intaglio, showing glabella, without the cheeks, and the nearly entire thorax and abdomen very well preserved.

Fig. 4.—A specimen in relievo, nearly entire, although not very sharply preserved. The glabella, which is partly removed, shows the hypostome (*h*) *in situ* beneath. Fig. 7 shows the same hypostome enlarged.

Fig. 5.—A group of probably four individuals, preserved in relief (with one pygidium in intaglio) in the same piece of matrix.

All the above are from the "Rottenstone band" in the Carboniferous Limestone of Matlock, Derbyshire, and are preserved in the British Museum (Nat. Hist.), Cromwell Road,

Fig. 6.—*Griffithides (Phillipsia) seminiferus*, Phil., sp. Outline restoration. Twice natural size.

Fig. 7.—The hypostome enlarged twice.

Fig. 8 *a*, *b*.—One of the thoracic somites. *a*, Seen from above; *b*, seen in section.

Fig. 9.—The pygidium restored in outline and enlarged.

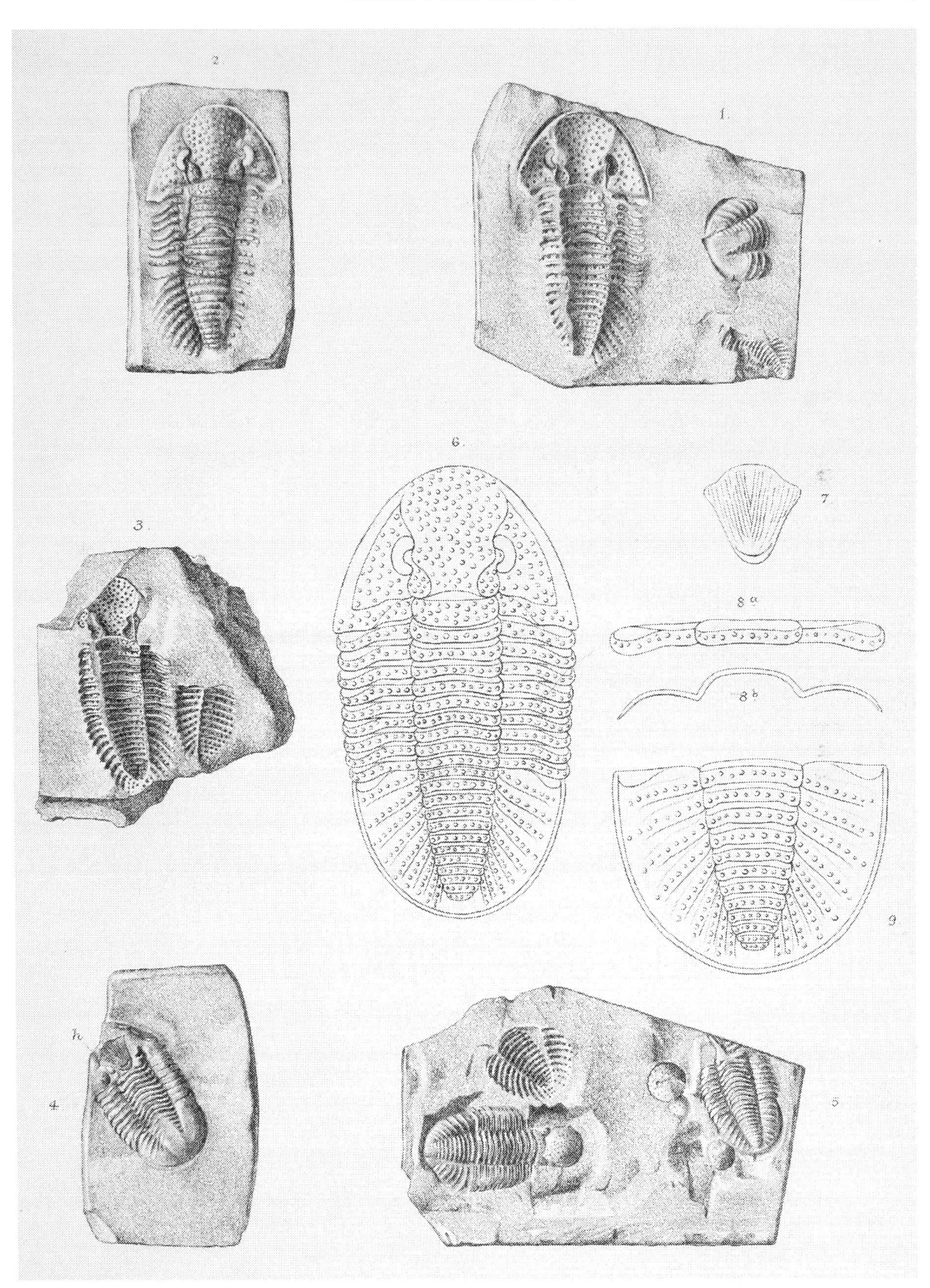

G. M. Woodward del. et lith.

West Newman & Co imp.

PLATE VI.

Carboniferous Trilobites.

Figs. 1 *a* and *b*.—*Griffithides globiceps*, Phillips, sp. *a*, Upper view; *b*, side view of an entire specimen from the Carboniferous Limestone, Millicent. Magnified once and a half natural size. Original specimen in the Museum of the Geological Survey of Ireland, Dublin. (P. 29.)

Fig. 2.—A nearly perfect head of *Griffithides acanthiceps*, H. Woodw., from the Carboniferous Limestone, Settle, Yorkshire. Enlarged one and a half times natural size. Original specimen in the Woodwardian Museum, Cambridge. (P. 32.)

Fig. 3.—An entire rolled-up example of *G. globiceps*. Enlarged twice natural size. From the same locality, formation, and collection as Fig. 1. (P. 29.)

Fig. 4.—Pygidium of same species, from same. (P. 29.)

Fig. 5.—Hypostome. Enlarged four times natural size. From the Carboniferous Limestone, Tyrone, Derryloran. Original in the Geological Survey Museum, Dublin. (P. 29.)

Fig. 6.—Pygidium from the Carboniferous Limestone, Bolland, Yorkshire. Magnified twice natural size. Original specimen in the British Museum (Nat. Hist.). (P. 29.)

Fig. 7.—*Griffithides longiceps*, Portlock. An entire specimen. Magnified twice natural size. Carboniferous Limestone. Ireland. Original in the Museum of Practical Geology, Jermyn Street. (P. 33.)

Fig. 8.—Outline restoration of same. Enlarged about four times. (P. 33.)

Fig. 9.—Pygidium of *G. longiceps*. Carboniferous Limestone; Ireland. Original specimen in the Museum of the Geological Survey of Ireland, Dublin. Magnified twice natural size. (P. 33.)

Fig. 10.—*G. acanthiceps*. Detached head from the Carboniferous Limestone of Settle, Yorkshire. Drawn natural size. Original in the Woodwardian Museum, Cambridge. (P. 32.)

Fig. 11.—*G. acanthiceps*. Detached head. Enlarged once and a half natural size. From the Carboniferous Limestone. Original in the cabinet of the Rev. E. O. de la Hey. (P. 32.)

Fig. 12.—*G. obsoletus*, Phil. Pygidium. Drawn natural size. From the Carboniferous Limestone of Bolland, Yorkshire. Original specimen in the British Museum (Nat. Hist.). (P. 35.)

Fig. 13.—*Griffithides platyceps*, Portlock (an imperfect glabella only). Enlarged twice natural size. From the Carboniferous Limestone, Tyrone, Derryloran. Original specimen in the Museum of the Geological Survey of Ireland, Dublin. (P. 34.)

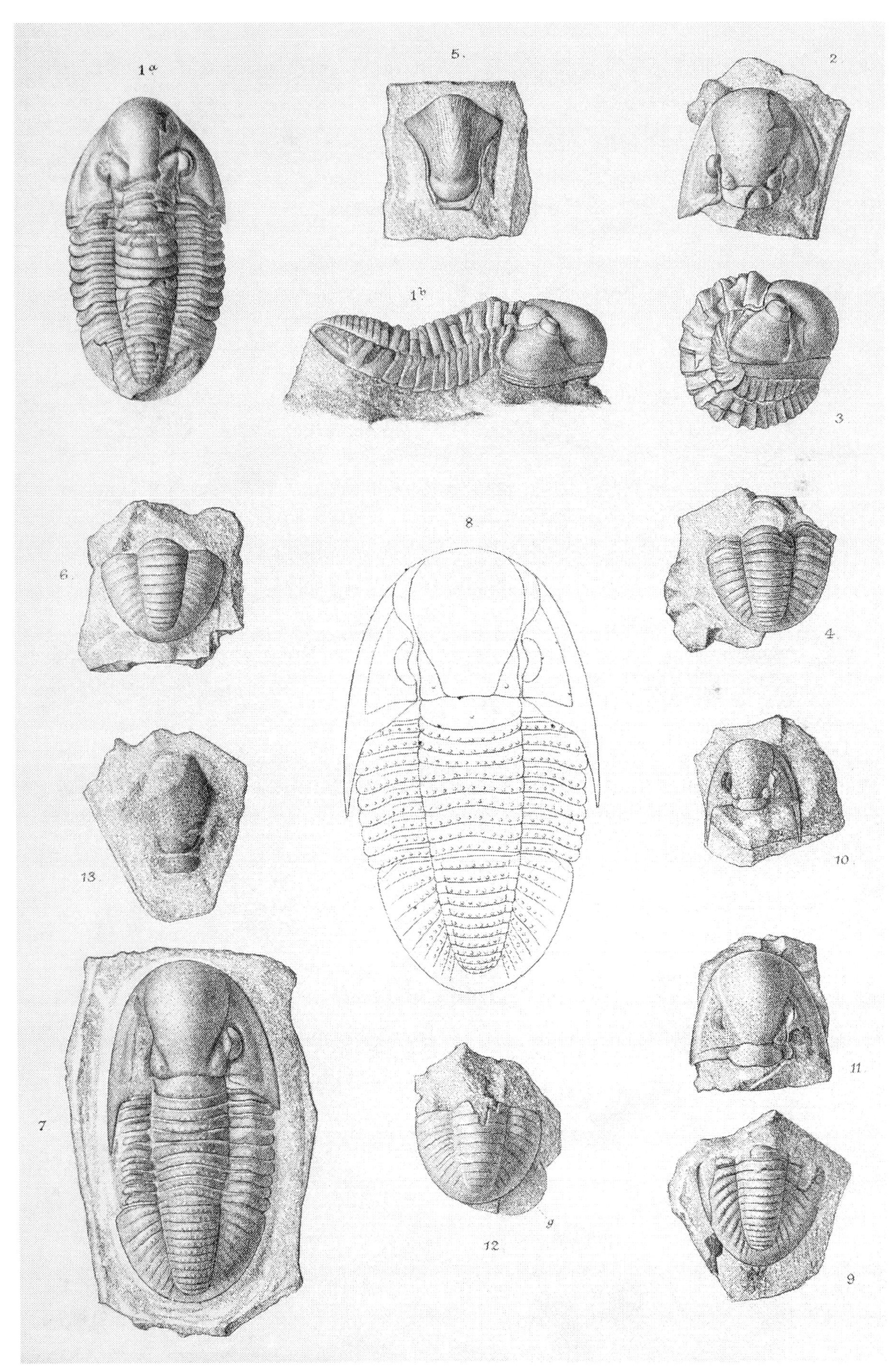

G. M. Woodward del. et lith.

West Newman & Co. imp.

PLATE VII.

CARBONIFEROUS TRILOBITES.

FIG. 1.—*Phillipsia quadrilimba*, Phil., sp. Pygidium. (Figure reproduced from Phillips' 'Geology of Yorkshire,' 1836, vol. ii, pl. xxii, fig. 2, p. 239.) Bolland, Yorkshire. (P. 26.)

FIG. 2.—*Griffithides acanthiceps*, H. Woodw. Nearly complete example, from Craco, near Grassington, Yorkshire. (P. 32.)

The original specimen in the cabinet of J. Aitken, Esq., Sandfield, Urmston, Manchester.

FIG. 3.—*Griffithides acanthiceps*.—Showing a detached head and pygidium from Settle, Yorkshire. (P. 33.)

The original specimen in the Woodwardian Museum, Cambridge.

FIG. 4.—*Phillipsia laticaudata*, H. Woodw. A detached glabella and pygidium from Bolland, Yorkshire. (P. 42.)

The original specimen in the Geological Department, British Museum (Natural History.)

FIGS. 5 *a*, *b*, *c*, and 6.—*Griffithides longispinus*, Portl.—5 *a*, profile; 5 *b*, front view of head; 5 *c*, view of pygidium of a rolled-up example (being the type-specimen figured by Portlock in his 'Geology of Londonderry and Tyrone,' pl. xxiv, fig. 12); 6, restoration of same unrolled. From Carnteel, Tyrone, Ireland. (P. 36.)

The original specimen in the Museum Practical Geology, Jermyn Street.

FIGS 7 and 8.—*Griffithides brevispinus*, H. Woodw. Portions of two heads. From Langside, Beith, Ayrshire. (P. 39.)

Original specimens in the collection of Mr. Robert Craig, Langside, Beith.

FIGS. 9, 10, 11, and 12.—*Griffithides moriceps*, H. Woodw. The heads only preserved. From Settle, Yorkshire. (P. 39.)

Original specimens in the Woodwardian Museum, Cambridge.

FIG. 13.—*Griffithides calcaratus*, M'Coy. Copied from Prof. M'Coy's figure in 'Synopsis Carbonif. Fossils of Ireland,' 1844, pl. iv, fig. 3. Roughan, Dungannon, Ireland. (P. 38.)

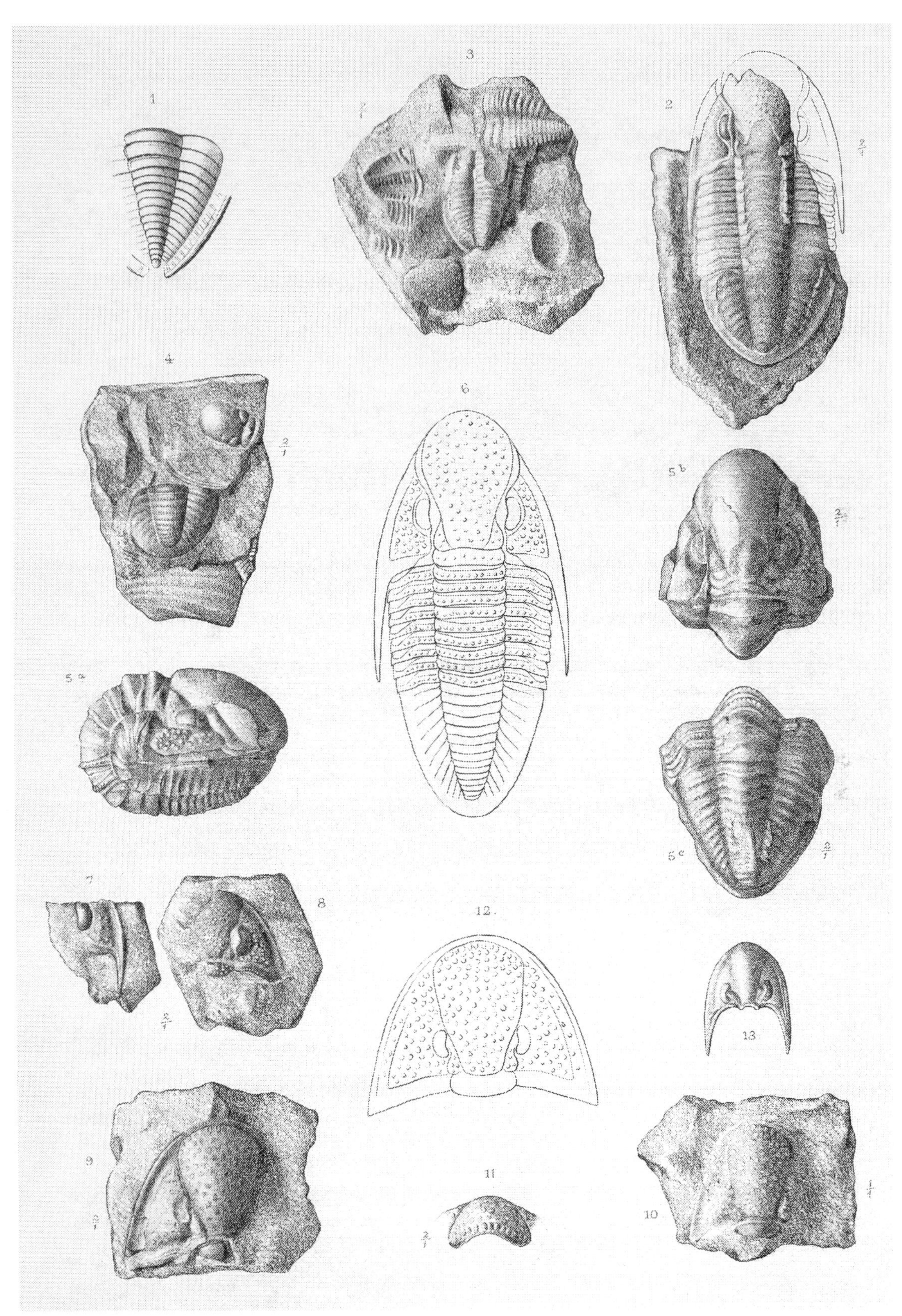

C. M. Woodward del. et lith.

West Newman & Co imp.

PLATE VIII.

CARBONIFEROUS TRILOBITES.

FIGS. 1—8.—*Brachymetopus ouralicus*, De Verneuil, sp., 1845. (P. 48.)

FIG. 1.—Head-shield, having both eyes perfect; enlarged four times natural size; from the Carboniferous Limestone, Settle, Yorkshire.

Original specimen in the Woodwardian Museum, Cambridge.

FIG. 2.—Another specimen, right eye wanting; neck-lobe well seen; enlarged four times natural size; from same locality and formation as preceding.

Original in the British Museum (Natural History) Cromwell Road.

FIG. 3.—A somewhat larger head-shield, enlarged three times natural size, having one of the cheek-spines perfect, but wanting the eyes; from the same locality and collection as Fig. 1.

FIGS. 4 and 5.—Two specimens, both with their head-shield and pygidium preserved close together, each in its own piece of matrix; enlarged twice the natural size; from the same locality, formation, and collection as figs. 1 and 3.

FIG. 6.—A very large pygidium, nearly perfect; magnified twice the natural size; from the same locality and collection as figs. 1, 3, 4, and 5.

FIG. 8.—Another pygidium, enlarged twice natural size; from the Carboniferous Limestone near Cork.

The original specimen in the collection of Joseph Wright, Esq., F.G.S., of Belfast.

FIG. 7.—Outline figure of *Brachymetopus ouralicus* restored and enlarged.

FIGS. 9—13.—BRACHYMETOPUS MACCOYI, Portlock, sp. (P. 52.)

FIG. 9.—A pygidium, enlarged three times, from the Black Carboniferous Limestone, Monaster.

The original specimen in the collection of Joseph Wright, Esq., F.G.S., of Belfast.

FIG. 10.—Head-shield of same from the same locality and collection as the above; enlarged three times.

FIG. 11.—Another head-shield of same, showing anterior margin of head well preserved, enlarged three times; from the Lower Limestone Shale (Carboniferous Limestone), County Limerick.

Original specimen in the Museum of the Geological Survey of Ireland, Dublin.

FIG. 12.—Head-shield having both cheek-spines preserved; enlarged three times natural size; from the same locality and collection as figs. 9 and 10.

FIG. 13.—Outline figure. Restored and enlarged of *B. Maccoyi.*

FIG. 14.—Pygidium, probably of *Griffithides seminiferus* (see Pl. V), from the Carboniferous Limestone, Settle, Yorkshire. (P. 28.)

Original specimen in the Woodwardian Museum, enlarged twice the natural size.

FIG. 15.—*Brachymetopus discors*, M'Coy. Pygidium, magnified six times, from the Carboniferous Limestone, Little Island, Cork. (P. 54.)

Original specimen from the collection of Joseph Wright, Esq., F.G.S., Belfast.

FIG. 16.—*Brachymetopus hibernicus*, H. Woodw., sp. Pygidium enlarged twice the natural size. From the Carboniferous Limestone, Kildare, Ireland. (P. 55.)

The original specimen in the British Museum (Natural History), Cromwell Road.

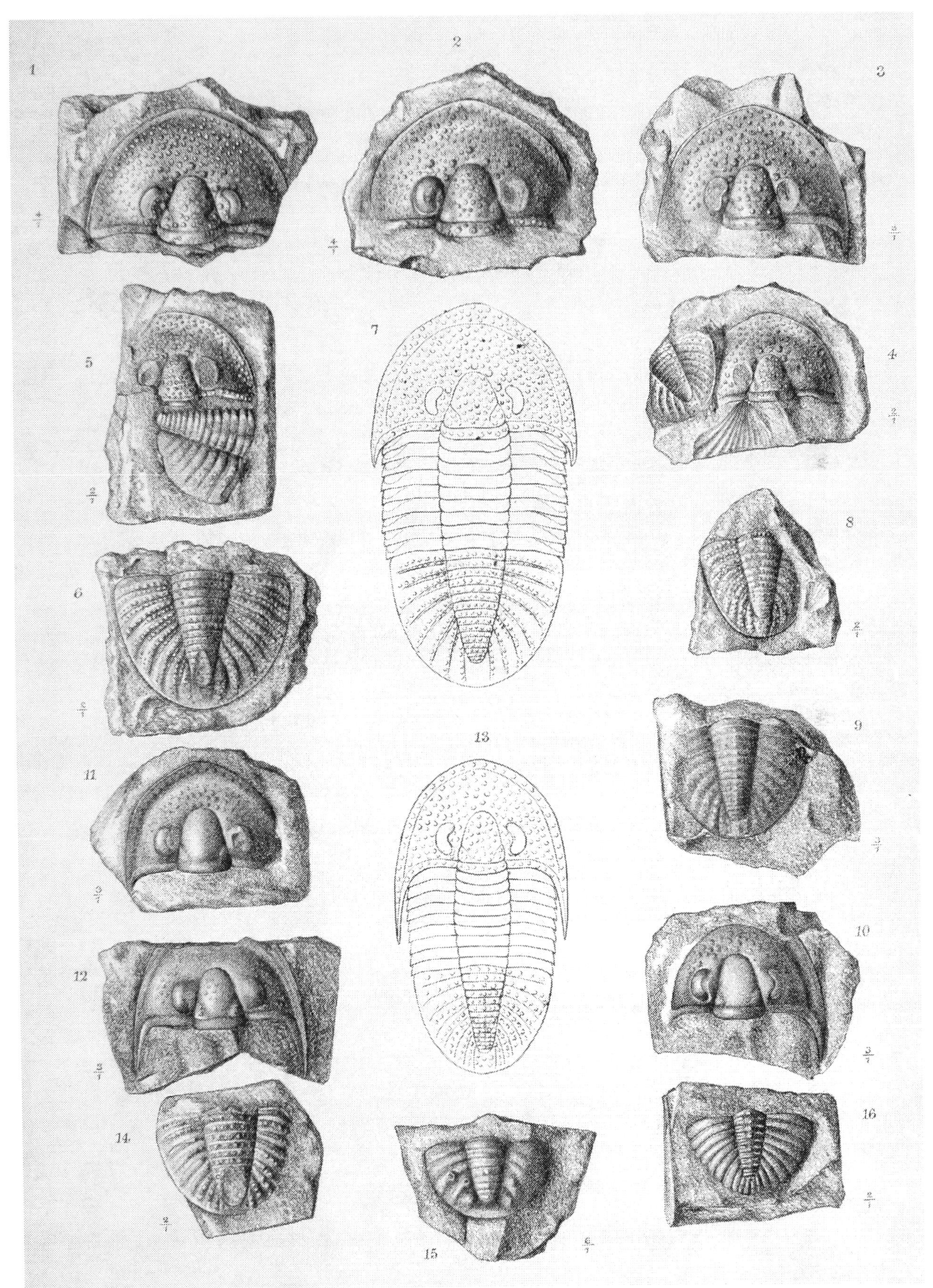

G. M. Woodward del. et lith.

West Newman & Co imp.

PLATE IX.

Carboniferous Trilobites and Recent Isopoda.

Fig. 1.—*Griffithides acanthiceps*, H. Woodw. Restoration in outline. (P. 32.) See Pl. VI, figs. 2, 10, and 11, and Pl. VII, figs. 2 and 3.

Fig. 2.—*Griffithides globiceps*, Phill., sp. Restoration in outline. (P. 29.) See Pl. VI, figs. 1, 3, 4, 5, and 6.

Fig. 3.—*Griffithides longispinus*, Portl. Glabella and hypostome (*h*), the former showing the preocular pore (*p*). Longnor, Derbyshire. See Pl. VII, figs. 5 and 6. (Pp. 36, 42.)

The original specimen in the Museum of Practical Geology.

Figs. 4 *a* and *b*.—*Griffithides glaber*, H. Woodw. 4 *a*, nearly entire specimen, enlarged twice nat. size; 4 *b*, body segments and tail, enlarged twice nat. size. From Castle Mumbles, Glamorganshire. (P. 40.)

The original specimens in the Museum of Practical Geology.

Figs. 5 *a* and *b*.—*Phillipsia scabra*, H. Woodw. 5 *a*, head-shield or cephalon, enlarged three times nat. size; 5 *b*, pygidium or abdomen, enlarged three times nat. size. From Vallis Vale, Frome, Somerset. (P. 43.)

The original specimens in the Museum of Practical Geology.

Figs. 6 *a* and *b*.—*Griffithides ? Carringtonensis*, Eth. MS. Two pygidia, enlarged three times nat. size. 6 *a*, Longnor, Derbyshire; 6 *b*, Narrowdale. (P. 41.)

The original specimens in the Museum of Practical Geology.

Fig. 7.—*Phillipsia carinata*, Salter, MS. Pygidium, enlarged three times nat. size. From the Carboniferous Limestone, Derbyshire. (P. 44.)

The original specimen in the Museum of Practical Geology.

Fig. 8.—*Serolis paradoxa* (recent). Enlarged one and a half nat. size. From Sandy Point, Straits of Magellan. Showing the pore (*p*) near the margin of the first thoracic somite. (See Appendix, p. 76.)

Fig. 9 *a*, *b*, *c*.—*Sphæroma gigas* (recent). Twice nat. size. From Kerguelen's Island. 9 *b*, front view of the head, showing pore (*p*); 9 *c*, half of same, enlarged four times, to show the pore more distinctly (*p*). (See Appendix, p. 76)

CARBONIFEROUS Plate IX.

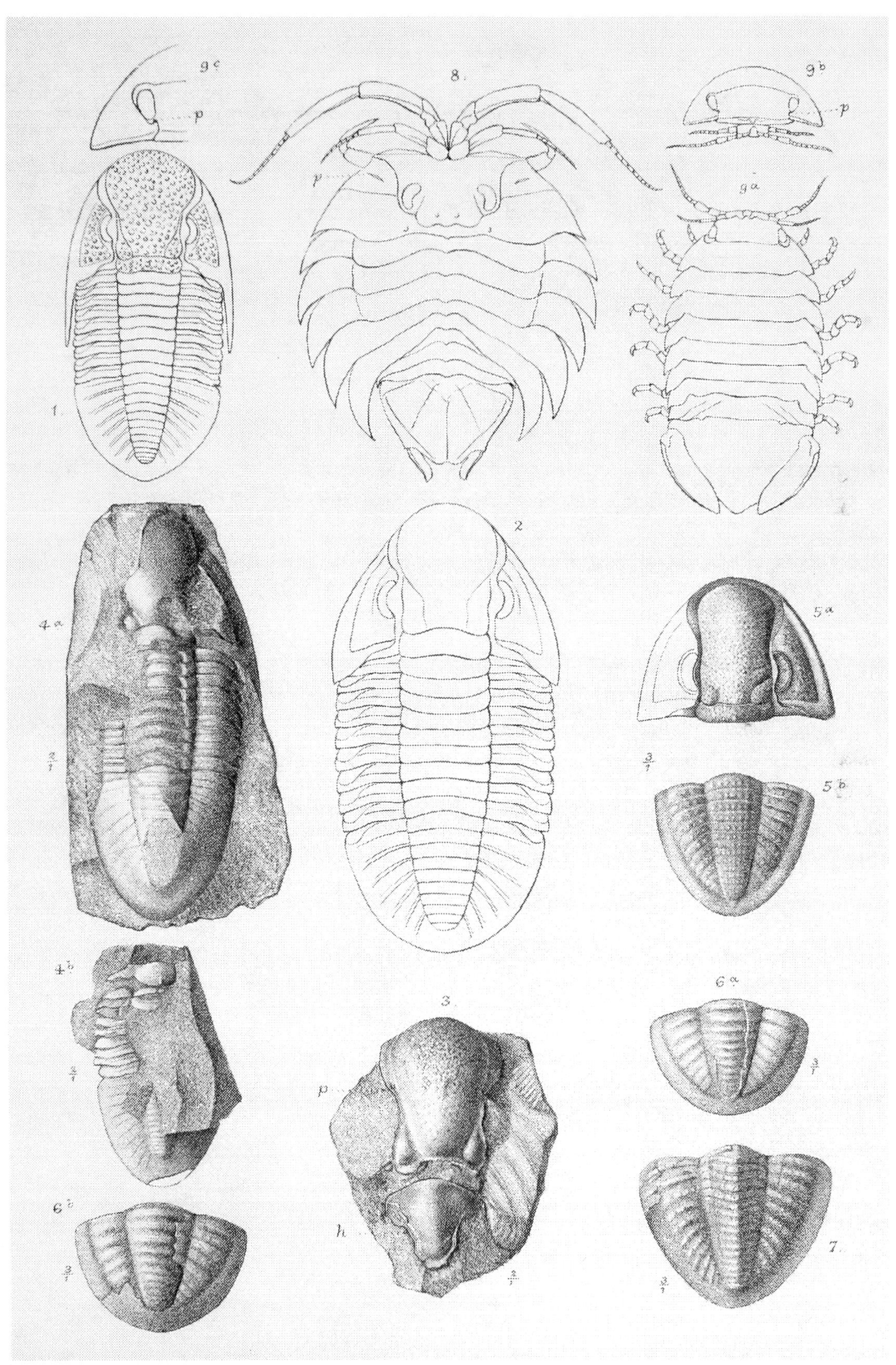

G. M. Woodward del. et lith.

West Newman & Co imp.

EXPLANATION OF PLATE X.

FIGS. 1—4.—*Phillipsia Leei*, H. Woodw., sp. nov. Lower (Carboniferous) Culm-shales, Waddon-Barton, near Chudleigh, Devonshire. (P. 66.)

FIG. 1.—Head-shield with part of the pygidium seen. The thoracic segments appear to underlie the head, and to be squeezed against the carapace beneath. Specimen enlarged three times.

FIG. 2.—Head-shield with five thoracic segments still united; also a detached pygidium. Enlarged three times.

FIG. 3.—A third head-shield with the left eye fairly preserved, showing also the facial sutures and the oblique furrows on the glabella. Enlarged three times.

FIG. 4.—Another head showing trace of the eye, and the facial suture and basal lobe of the glabella; also part of thoracic segments. Enlarged four times.

FIGS. 5, 6 *a*, *b*—8, 8 *a*.—*Phillipsia minor*, H. Woodw., sp. nov. Formation and locality the same as that of the preceding species. (P. 68.)

FIG. 5.—A pygidium seen in intaglio. Enlarged four times.

FIG. 6 *a* and *b* only.—*a*. Head-shield; and *b*, pygidium of same. Enlarged three times natural size.

FIG. 7 *a*.—A nearly entire example, seen in intaglio, enlarged eight times natural size, with one detached free-cheek (*c*) and hypostome (*h*).

FIG. 8 *a* only.—A small detached head. Enlarged three times natural size.

FIG. 8 *b*, and figs. 9—12.—*Phillipsia Cliffordi*, H. Woodw., sp. nov. Formation and locality the same as that of the preceding species. (P. 69.)

FIG. 8 *b* only.—Pygidium. Enlarged three times.

FIGS. 9 and 10.—Pygidia. do.

FIG. 11.—Portion of a head-shield showing the right side of the glabella with the free-cheek attached, and a detached pygidium. Enlarged four times.

FIG. 12.—A detached pygidium. Enlarged eight times.

FIGS. 6 *c*, *d* (only) and FIG. 13.—*Phillipsia articulosa*, H. Woodw., sp. nov. Formation and locality the same as that of the preceding species. (P. 70.)

FIG. 6.—Two pygidia (*c*) seen from beneath, (*d*) seen from above. Enlarged three times natural size.

FIG. 13.—Detached pygidium. Enlarged four times.

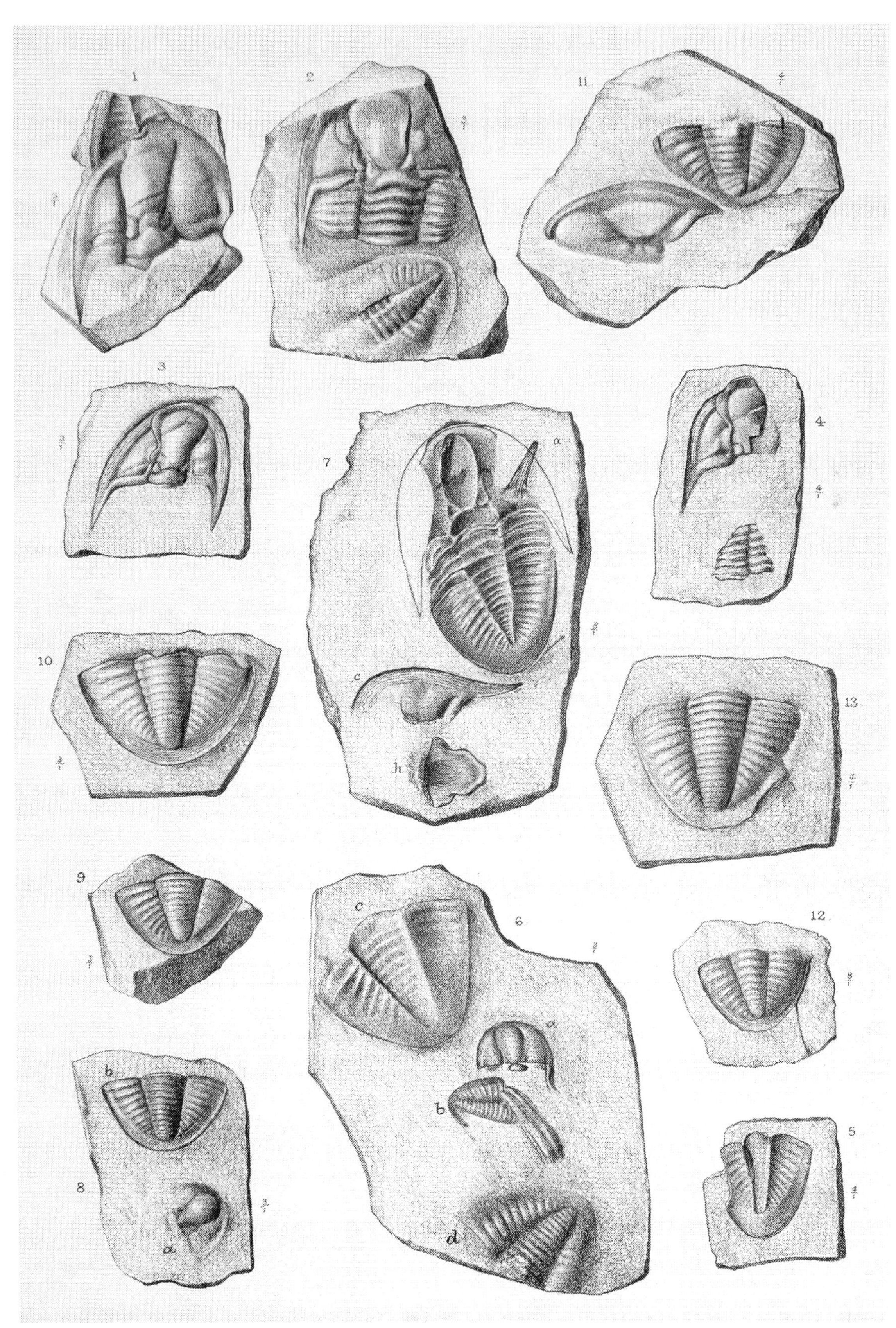

G. M. Woodward del. et lith.

West Newman & C° imp.

Printed in the United States
By Bookmasters